Second-Rank Cities in Europe

Second-rank cities are back on the academic scene, capturing the interest of scholars with their unexpected recent performance with respect to first-rank cities. Looking at the data on average urban GDP growth in 139 European cities since 1996, the relatively strong position of large cities (over 1.5 million inhabitants) on national growth coincides with the periods of fastest expansion, while at times of slowdown second-rank cities prevail. However, in the recent period of economic downturn, second-rank cities have recorded annual GDP growth rates far less negative than those of capital cities; and in some European countries, like Austria and Germany, all cities have outperformed their capitals.

An intringuing explanation of the dynamics of second-rank cities has been based on agglomeration economy theories. However, merely linking agglomeration economies to urban size in order to interpret urban performance is neither convincing nor sufficient, and it calls for additional investigation into how agglomeration economies work. This volume claims that interpretation of the current dynamics in European urban systems – especially in the western part of Europe – would benefit from exploitation of the traditional concept of agglomeration economies. However, necessary for this purpose are more in-depth considerations on the nature, scope, intensity and causes of agglomeration economies, which do not relate their existence solely to urban size. And this is where the main challenge for scholars lies – in the interpretation of the missing link between agglomeration economies and urban dynamics. This book was originally published as a special issue of *European Planning Studies*.

Roberto Camagni is Professor of Urban Economics at the Politecnico di Milano, Italy. He is a past president of the European Regional Science Association and of GREMI. He is the author of many scientific papers and a textbook on Urban Economics, published in Italian, French and Spanish.

Roberta Capello is Professor in Regional Economics at the Politecnico di Milano, Italy. She is a past president of the Regional Science Association International. She is the editor in chief of *Papers in Regional Science* and of the *Italian Journal of Regional Science*. She is the author of many scientific papers and the editor of *Territorial Patterns of Innovation: An Inquiry on the Knowledge Economy in European Regions* (Routledge, 2013) and of a textbook on Regional Economics (Routledge, 2007, second edition 2016) published in Italian, English and Chinese.

Second-Rank Cities in Europe

Structural dynamics and growth potential

Edited by
Roberto Camagni and Roberta Capello

LONDON AND NEW YORK

First published 2016
by Routledge
2 Park Square, Milton Park, Abingdon, Oxon, OX14 4RN, UK

and by Routledge
711 Third Avenue, New York, NY 10017, USA

First issued in paperback 2017

Routledge is an imprint of the Taylor & Francis Group, an informa business

© 2016 Taylor & Francis

British Library Cataloguing in Publication Data
A catalogue record for this book is available from the British Library

ISBN 13: 978-1-138-29501-8 (pbk)
ISBN 13: 978-1-138-95104-4 (hbk)

Typeset in Times New Roman
by diacriTech, Chennai

Publisher's Note
The publisher accepts responsibility for any inconsistencies that may have arisen during the conversion of this book from journal articles to book chapters, namely the possible inclusion of journal terminology.

Disclaimer
Every effort has been made to contact copyright holders for their permission to reprint material in this book. The publishers would be grateful to hear from any copyright holder who is not here acknowledged and will undertake to rectify any errors or omissions in future editions of this book.

Contents

CONTENTS

Citation Information

The chapters in this book were originally published in *European Planning Studies*, volume 23, issue 6 (June 2015). When citing this material, please use the original page numbering for each article, as follows:

Introduction
Second-Rank City Dynamics: Theoretical Interpretations Behind Their Growth Potentials
Roberto Camagni and Roberta Capello
European Planning Studies, volume 23, issue 6 (June 2015) pp. 1041–1053

Chapter 1
City Size and Economic Performance: Is Bigger Better, Small More Beautiful or Middling Marvellous?
Michael Parkinson, Richard Meegan and Jay Karecha
European Planning Studies, volume 23, issue 6 (June 2015) pp. 1054–1068

Chapter 2
The Rise of Second-Rank Cities: What Role for Agglomeration Economies?
Roberto Camagni, Roberta Capello and Andrea Caragliu
European Planning Studies, volume 23, issue 6 (June 2015) pp. 1069–1089

Chapter 3
Borrowed Size, Agglomeration Shadows and Cultural Amenities in North-West Europe
Martijn J. Burger, Evert J. Meijers, Marloes M. Hoogerbrugge and Jaume Masip Tresserra
European Planning Studies, volume 23, issue 6 (June 2015) pp. 1090–1109

Chapter 4
Related Variety and Regional Economic Growth in a Cross-Section of European Urban Regions
Frank van Oort, Stefan de Geus and Teodora Dogaru
European Planning Studies, volume 23, issue 6 (June 2015) pp. 1110–1127

Chapter 5
Assessing Polycentric Urban Systems in the OECD: Country, Regional and Metropolitan Perspectives
Monica Brezzi and Paolo Veneri
European Planning Studies, volume 23, issue 6 (June 2015) pp. 1128–1145

Chapter 6

First- and Second-Tier Cities in Regional Agglomeration Models
Chiara Agnoletti, Chiara Bocci, Sabrina Iommi, Patrizia Lattarulo and Donatella Marinari
European Planning Studies, volume 23, issue 6 (June 2015) pp. 1146–1168

Chapter 7

Polycentric Metropolitan Development: From Structural Assessment to Processual Dimensions
Rudolf Giffinger and Johannes Suitner
European Planning Studies, volume 23, issue 6 (June 2015) pp. 1169–1186

For any permissions-related enquiries please visit
http://www.tandfonline.com/page/help/permissions

Notes on Contributors

Chiara Agnoletti works for the Institute of Regional Economic Planning for Tuscany, based in Florence, Italy.

Chiara Bocci works for the Institute of Regional Economic Planning for Tuscany, based in Florence, Italy.

Monica Brezzi is Head of the Regional Analysis and Statistics Unit in the OECD Directorate for Public Governance and Territorial Development. Her current activities focus on the analysis of regional comparative advantages and the assessment of policies to reduce inequalities in the access to key services for citizens.

Martijn J. Burger is a Research Fellow in the Tinbergen Institute and Assistant Professor of Industrial and Regional Economics at Erasmus University Rotterdam, the Netherlands. His research interests include happiness economics, medical sociology, health studies and spatial epidemiology.

Roberto Camagni is Professor of Urban Economics at the Politecnico di Milano, Italy. He is a past president of the European Regional Science Association and of GREMI. He is the author of many scientific papers and a textbook on Urban Economics, published in Italian, French and Spanish.

Roberta Capello is Professor in Regional Economics at the Politecnico di Milano, Italy. She is a past president of the Regional Science Association International. She is the editor in chief of *Papers in Regional Science* and of the *Italian Journal of Regional Science*. She is the author of many scientific papers and the author of a textbook on Regional Economics published in Italian, English (Routledge, 2007, second edition 2016) and Chinese.

Andrea Caragliu is Assistant Professor of Regional and Urban Economics at Politecnico di Milano, Italy. He has published articles in *Regional Studies*, *International Regional Science Review,* and *Regional Science Policy & Practice*.

Stefan de Geus works in Business Development and Investor Relations at Troidos Investment Management in Zeist, the Netherlands.

Teodora Dogaru is based in the Department of Economic Analysis and Business Administration at the University of A Coruña, Spain, and the Faculty of Geosciences at Utrecht University, the Netherlands.

NOTES ON CONTRIBUTORS

Rudolf Giffinger is Professor in Regional Science and Head of the Centre of Regional Science, at the Vienna University of Technology, Austria. His research mainly concentrates on intra-urban development, urban decay and segregation and integration, as well as on urban/metropolitan competitiveness of selected cities and respective strategic issues.

Marloes M. Hoogerbrugge is a Junior Researcher in the OTB Research Institute for Housing, Urban and Mobility Studies at Delft University of Technology, the Netherlands.

Sabrina Iommi works for the Institute of Regional Economic Planning for Tuscany, based in Florence, Italy.

Jay Karecha is a Research Fellow and Lead Team Member for Quantitative Analysis in the European Institute for Urban Affairs at Liverpool John Moores University, Liverpool, UK. He is currently carrying out data analysis for the ESPON Secondary Growth Poles project, examining the social and economic performance of European cities.

Patrizia Lattarulo is the head of the "Public Economics and Territory" research area of the Institute of Regional Economic Planning for Tuscany, based in Florence, Italy.

Donatella Marinari works for the Institute of Regional Economic Planning for Tuscany, based in Florence, Italy.

Richard Meegan is Professor of Economic Geography in the European Institute for Urban Affairs at Liverpool John Moores University, Liverpool, UK.

Evert J. Meijers is a Researcher in the OTB Research Institute for Housing, Urban and Mobility Studies at Delft University of Technology, the Netherlands. His main research interests include themes such as urban competitiveness, regional development, co-operation between cities and urban governance and planning.

Michael Parkinson is Director of the European Institute for Urban Affairs at Liverpool John Moores University, Liverpool, UK. He has acted as adviser on urban affairs to the European Commission, OECD, EUROCITIES, the Department of Communities and Local Government, the National Audit Office, the House of Commons Select Committees, the Core Cities and a range of cities in the UK.

Johannes Suitner is a Post-Doctoral Researcher in Urban Studies at Vienna University of Technology, Vienna, Austria.

Jaume Masip Tresserra is a PhD candidate in the Faculty of Architecture and the Built Environment at Delft University of Technology, the Netherlands. His main research asks if a polycentric structure at intra-metropolitan scale positively associates with increasing the economic, social and environmental performance of metropolitan regions.

Frank van Oort is Professor of Human Geography and Planning at Utrecht University, the Netherlands. He is the editor of the *Regional Studies* journal.

Paolo Veneri is an Economist at the Organisation for Economic Co-operation and Development (OECD), carrying out research on urban and regional issues. He is the main author of an OECD report entitled 'Rural-urban partnership: an integrated approach to economic development', 9789264204805.

INTRODUCTION

Second-Rank City Dynamics: Theoretical Interpretations Behind Their Growth Potentials

ROBERTO CAMAGNI & ROBERTA CAPELLO

Politecnico di Milano, Milano, Italy

1. The Missing Link Between Agglomeration Economies and Urban Dynamics: Theme of the Special Issue

Second-rank cities are back on the academic scene, capturing the interest of scholars with their unexpected recent performance with respect to first-rank cities. In the data on average urban GDP growth in 139 European cities since 1996, the relatively strong position of large cities (over 1.5 million inhabitants) in national growth coincides with periods of fastest expansion, while at times of slowdown second-rank cities prevail (Camagni et al., 2014a). Especially in the recent period of economic downturn, second-rank cities have recorded annual GDP growth rates that are much less negative than those of capital cities; and in some European countries, like Austria and Germany, all cities have outperformed their capitals (Parkinson et al., 2014). This trend is not confined to Europe alone. In the USA, between 1969 and 2007, the largest eight Metropolitan Statistical Areas (now on MSAs) grew by only one-third of the rate of the other three smaller MSA categories. Moreover, after 1990, the performance of the large MSAs was only slightly better, since their growth rates were still only about one-half of the average rates for the three smaller MSA size categories. The Big MSAs' growth rate barely exceeded those of non-metropolitan areas: indeed, they even trailed the average growth rate of non-metropolitan areas (Partridge, 2010).

This is not the first time that the dynamics of second-rank cities have captured the interest of scholars; over time, different interpretations have been put forward of their relatively good performance. All of them are interesting, but none of them is fully convincing because they are linked more to cyclical than structural factors. During the 1980s, the fast growth of medium-sized cities in developing countries was explained as the conse-

1

quence of repulsive forces from primate cities (Azzoni, 1986; Richardson, 1980; Townroe & Kean, 1984). If this may have been an explanation in a particular period of time for emerging countries, it cannot be generalized to all periods and all countries.

At the end of the 1990s, in their book titled *Second Tier Cities: Rapid Growth Beyond the Metropolis*, Markusen et al. (1999, p. 3) once again asked the question "why newer, smaller cities had grown at the expense of older, larger ones, upsetting urban hierarchies". At that time, the reply was found in the role that second-rank cities were playing in the new industrial organization. They were identified as distinct areas of economic activity, constituting separate labour markets, where a specialized set of trade-oriented industries took root and flourished to establish employment and population-growth trajectories: the so-called new industrial spaces. Their success was strongly linked to the need of some industries to exploit localization (or district) economies, more than to the urbanization economies typical of large cities. Also this interpretation of second-rank city dynamics had more to do with a particular stage of industrial development; but it could not hold for all small- and medium-sized cities, in all countries and periods.

The most recent interpretation of the growth of second-rank cities has been rather different. It draws on the tradition of urban economic theory dealing with increasing/decreasing returns to urban size. In the recent European urban dynamics, in fact, the decline of large cities has been attributed to the emergence of decreasing returns to urban size, while second-rank cities have been interpreted as enjoying increasing returns, although they are far from the optimal city size envisaged by urban economic theory (Dijkstra *et al.*, 2013). Agglomeration economies have been reprised as the main explanation for urban performance interpreted as a law of structural dynamics able to explain success and failure of cities.

Linking the explanation of urban dynamics to agglomeration theories seems the most interesting approach. However, as presented in Dijkstra *et al.*, 2013, the mere linking of agglomeration economies to urban size in order to interpret urban performance is neither convincing nor sufficient, and calls for additional investigation of how agglomeration economies work. In such an approach, in fact, it is not clear why in certain periods of time large cities enter a period of decreasing returns, while in others they enjoy increasing returns. By the same token, it is not clear why, at a certain point in time, second-rank cities start enjoying agglomeration economies. This issue mixes two problems together: a structural and a cyclical one. The former concerns the existence of a theoretical explanation for how cities of different sizes can exploit increasing returns to urban scale (the "how" problem); the latter problem concerns the question of whether cities of different sizes are better suited for expansion or crisis periods, or for periods of emerging technological/organizational paradigms or periods of paradigm diffusion (the "when" question). The "how" issue is logically propedeutical to, and more interesting than, the "when" issue, because descriptive evidence shows that there are some large cities able to play a role in their national economies in all periods, and some second-rank cities that still lag behind, suggesting that, for the same urban size, agglomeration economies play a positive role for some cities and not for others (Camagni *et al.*, 2014b).

Our impression is that interpretation of the current dynamics in European urban systems—especially in the western part of Europe—would benefit from exploitation of the traditional concept of agglomeration economies. However, necessary for this purpose are more in-depth considerations on the nature, scope, intensity and causes of agglomeration economies which do not relate their existence solely to urban size. And this is where the main challenge for scholars lies: interpretation of the *missing link between agglomeration economies and urban dynamics* (Meijers, 2013).

The main aim of this special issue is to take up this challenge and to present the most advanced efforts made to interpret second-rank city dynamics and its role in the evolution of polycentric urban systems, by improving on existing agglomeration theories. In this way, the special issue contributes to the most advanced debate on how to link city and urban structure evolution to agglomeration economies.

2. Recent Theoretical Interpretations of the Missing Link: A Common Effort by Different Approaches

While the dynamics of "world cities" (Friedmann, 1986), "global cities" (Sassen, 1991), or "global-city regions" (Scott, 2001) was for long interpreted as stemming from their capacity to attract high-value functions, offering them a mix of talents and skills in a broad range of specialized fields, nowadays this seems to be no longer the case. Large cities do not grow faster than small- and medium-sized ones, and the current dynamics of European urban systems cannot be interpreted through the advantages of urban size. Empirical analyses of urban system dynamics, in fact, have shown that there are no regularities between the size and growth of cities, and they stress the need for an explanation of the missing link between agglomeration economies and urban dynamics.

The literature on agglomeration economies has traditionally highlighted three aspects that are inherently part of this concept: indivisibilities, synergies and physical proximity (Capello, 2009). Indivisibilities occur when the scale of agglomerated activities adds to productivity by causing shifts in a firm's production or cost curve (Cohen *et al.*, 2009; Rosenthal & Strange, 2001), but also shifts in aggregate urban efficiency that allow quantum jumps (leaps) in the infrastructure system. Indivisibilities prevail when an industrial perspective is taken: some sectors are more dependent on large-scale production processes than others, and some sectors derive many advantages from the presence of other sectors generating efficient and large "industrial complexes" (Isard & Schooler, 1959). In light of these sectoral peculiarities, a large body of literature measures the extent to which the presence of a mix of industries or of a single industry generates greater agglomeration advantages (Carlino, 1980; Henderson, 1985). Synergies relate to the socio-cultural dimension: trust, sense of belonging, cultural and religious homogeneity, these being typical features of agglomerated and specialized areas. They heighten the intensity of local market and non-market interactions, thus giving rise to increasing returns on production factors via transaction/production cost minimization for a given output (Becattini, 1989) or innovation-enhancing processes (Camagni, 1991; Storper, 1995). Proximity is by definition linked to the geographical dimension of agglomeration and interaction effects: if information and transportation costs were nil, in the absence of scale economies there would be no reason to concentrate activities because doing so would not produce "economies". In this sense, agglomeration economies are "proximity economies".

These three elements explain the differences among the approaches to the sources of agglomeration economies, their nature, scope and intensity being explained from single perspectives: technical scale effects, easier market interactions, or limited distance friction. If one looks carefully into the most recent literature, a striking aspect emerges: whatever the approach taken, the most updated agglomeration theories undertake the explanation of urban dynamics through the concept of agglomeration economies by directly or indirectly taking up the challenge of explaining the missing link.

Table 1 sketches the new theories pertaining to the different approaches: the search for micro-economic foundations of agglomeration economies; the wider search for spatial interaction constituting the geographical foundations of agglomeration economies; the search for the macro-territorial foundations of agglomeration economies typical of recent regional economics contributions. As we see below, these approaches are rather different in nature. They capture different and complementary aspects of agglomeration economies, and stress particular elements that pertain to the foundations of agglomeration economies. Indivisibilities (labour-market indivisibilities, production indivisibilities) prevail in the micro-economic approach, allowing the reduction of production costs; physical proximity is the domain of the geographical approach; while synergy and interactions (limiting transaction costs) prevail in a macro-territorial approach. Notwithstanding these differences, common to all the approaches is the attempt to directly or indirectly explain urban dynamics through agglomeration economies, and they call for additional reflection on the concept itself of agglomeration economies in its traditional form. Advanced theoretical efforts already exist in all types of approaches applied to the dynamics of single cities, as well as to the evolution of the entire urban system, like regional and national polycentric urban structures or city networks.

3. Micro-Economic Foundations of Agglomeration Economies: New Perspectives

The early studies on agglomeration economies sought to explain theoretically and test empirically whether the scale of agglomerated activities added to productivity. The

Table 1.   Interpretations of the missing link between agglomeration economies and urban dynamics

	Micro-economic foundations	Geographical foundations	Macro-territorial foundations
Sources of advantages internal to the city	Heterogeneous Industries Industry mix	Milieu effects	Quality of local factors, of activities hosted, of urban infrastructure
Sources of advantages internal to the urban system	Agglomeration shadows (Growth shadows)	Borrowed size Externality fields	Density and quality of external linkages City networks (Network externalities)
Interpretation of the link	Specificities of single industries Spatial competition effects	A regionalization of agglomeration externalities	Quality of territorial capital assets
Interpretative logics: from size to growth through efficiency	Size ↓ Static efficiency ↓ Competitiveness ↓ Physical growth	Size + proximity ↓ Externalities ↓ Competitiveness ↓ Physical growth	Size + territorial capital ↓ Dynamic efficiency ↓ Innovation ↓ Physical growth and structural evolution

well-known dichotomy drawn between urbanization and localization economies reflected these attempts, see among others (Carlino, 1980; Henderson, 1985; Hoch, 1972; Mera, 1973; Mills, 1970; Moomaw, 1983; Segal, 1976; Shefer, 1973; Sweiskauskas, 1975). The analytical framework used by this approach was in fact the identification of whether scale economies are related to the scale of the local specialization industry or to diversification and cross-fertilization among industries. Nevertheless, no consensus was achieved even when geo-referenced data on establishments and advanced spatial econometric techniques opened the way to more sophisticated analyses[1] and still today many studies try to find the definitive answer to the question "who is right: Marshall or Jacobs?" (quoted from the title of Beaudry & Schiffauerova, 2009).

In this approach, the sources of agglomeration economies are micro-economic in nature. They are identified in production cost minimization in accordance with the typical Marshallian tradition: input sharing interpreted in terms of increasing returns to scale in production; the local labour market as the size which allows a better match between employers' needs and workers' skills and reduced risks for both; home market effects and scale of pecuniary externalities (Krugman, 1991); superior technical knowledge achieved thanks to a large scale of operation; and highly specialized local labour markets (Rosenthal & Strange, 2001).

The novelty in this field consists in investigation of the different ways in which the three well-known Marshallian forces—input sharing, market pooling and knowledge spillovers—play a role in agglomeration and locational choices according to the specificities of single industries (Faggio et al., 2013). Heterogeneity across industries seems pervasive, and the search for a "universal understanding of micro-economic foundations applying to all industries appears to be somewhat visionary and misleading" (Faggio et al., 2013, p. 4).

Even if not directly, this approach explains urban dynamics by considering the forces behind industry co-agglomeration in specific metropolitan areas. This approach suggests that it is not the size of a city, but rather the presence of the right combination of micro-economic elements, that explains its dynamics.

When considering the formation of large urban systems, the micro-economic approach focuses on the *shadow effects* that the presence of large cities exerts on surrounding smaller cities (Table 1). In their evolutionary model, Fujita and Mori (1996, 1997) assume the existence of a single city in which a variety of goods is produced, and a hinterland which produces agricultural goods. The presence of a population in the hinterland becomes of interest to entrepreneurs seeking market opportunities, and this is sufficient to develop the city. The overall population reaches a critical threshold at which equilibrium becomes unstable (Dobkins & Ioannides, 2001), and a catastrophic bifurcation towards a duocentric system may occur. The location in which the new city is formed is not random; a core prediction of standard new economic geography is that the existence of *shadow economies* (Krugman, 1993) prevents urban areas from arising too close to other urban areas of equal or larger size owing to fierce spatial price competition. This is in line with the static perspective of the central place theory (Christaller, 1933).

According to these new perspectives, smaller cities have lower growth possibilities when located close to large cities. However, current urban systems seem highly diversified: they sometimes exhibit the existence of agglomeration shadows and sometimes oppose even positive effects of large cities on smaller ones (Partridge et al., 2009). Better interpretation of the reality requires additional theoretical insights into the geographical foundations of agglomeration economies.

4. Geographical Foundations of Agglomeration Economies: New Perspectives

A different and complementary approach to the interpretation of agglomeration economies and urban dynamics looks for the so-called *geographical foundations* of agglomeration economies (Burger *et al.*, 2014; Meijers, 2013) (Table 1). Traditionally, this approach has concerned itself with the proximity and non-market interactions among firms and people that give rise to "district" and "milieu" effects. These effects generate growth on the basis of both static and dynamic efficiency elements (Camagni, 1991). In more recent studies, this approach has highlighted the fact that urban agglomeration effects are not necessarily confined to the physical boundaries of a city but spill over to surrounding areas.

The starting point of this last approach is the concept of "borrowed size" developed by Alonso (1973); " ... a small city or a metropolitan area exhibits some of the characteristics of a larger one if it is near other population concentrations" (Alonso, 1973, p. 200). Behind this statement lies the claim that smaller places can "borrow" some of the agglomeration benefits of their large neighbours, while avoiding agglomeration costs.[2]

The physical distance at which agglomeration economies are able to exert their effects is the main element in this approach, which explains why smaller cities can sometimes grow thanks to (and at the expense of) nearby large cities. This approach can easily explain why smaller cities can grow more than larger cities, as well as why efficient polycentric urban structures at the local (regional) level exist (Brezzi & Veneri, 2014; Agnoletti *et al.*, 2014; Giffinger and Suitner, 2014). The concepts of "externality fields" (Phelps *et al.*, 2001) or "regional externalities" (Parr, 2002) have been proposed to highlight the spatial coverage of urban advantages extending far beyond the city's boundaries.

The capacity of some specific second-rank cities to grow more than large cities is interpreted as stemming from the prevalence of "borrowed size" over "shadow economies". As Burger *et al.* claim, the strategic aspect to understand is which cities are advantaged and which cities are disadvantaged by spatial interdependencies. In other words, the question is which cities benefit from borrowed size and which face agglomeration shadows. The reply is complicated, and probably varies according to the urban function analysed. In the case of cultural amenities (museums, art galleries, etc.), Burger *et al.* (2014) make a first attempt to measure if and when borrowed size prevails over agglomeration shadows. They consider the sizes of cities and their positions in the urban hierarchy: larger cities within a functional urban area obtain the highest benefits from the size of the rest of the area in which they are located and from access to a large (inter)national market. Lower order places are less likely to borrow size from other places: they remain in the shadow of a first-rank city and undergo spatial competition effects. From this perspective, therefore, the dynamics of second-rank cities can be better explained when a regionalization of urban externalities is conceptualized.

5. Macro-Territorial Foundations of Agglomeration Economies: New Perspectives

The last approach considered investigates the macro-territorial foundations of agglomeration economies, taking the city as the main unit of analysis, as do the large number of empirical studies measuring the scale effects of urban size (Henderson, 1974, 1985). In this tradition, the success of cities is attributed to the existence of agglomeration economies; conversely, urban decline is explained by the loss of increasing returns when a city reaches an excessive size.

Recently, the idea that one single optimal city size exists has been abandoned; and so too has the opposite, too simplistic, view that infinite optimal city sizes exist, one for each city. A first theoretical solution resides in acknowledgement that other measurable factors affecting urban costs and benefits contribute together with pure size to the equilibrium size of the city (Camagni *et al.*, 2013). A second solution starts from commonsense acknowledgement of the existence of different size classes of cities (small, medium, large) and the consideration that each class encompasses structurally similar cities (Camagni *et al.*, 2014b). In fact, large cities are de-specialized in terms of activity sectors, and they host high-level functions and occupations; medium-sized cities are generally more specialized and high-performing in the specialization sectors; small cities mainly host low-level skills and activities (Conti & Dematteis, 1995).

In a simplified view, agglomeration economies may be taken for granted in small- and medium-sized cities; and only in large cities should the problem of a downturn in urban returns to scale eventually emerge. Assuming a more complex view, a new theoretical conjecture claims that the exploitation of agglomeration economies is relatively smooth within the three/four traditional size classes (small, medium and large cities), but it implies the presence of specific limiting/enabling factors when cities approach some critical instability point (Camagni *et al.*, 2014b). Therefore, cities may experience a halt in their growth path, and even a decline, irrespective of their size class in the absence of these conditioning factors. These factors are not really quantitative in nature; rather, they are qualitative, and quantum jumps in their endowment are needed at specific intervals if agglomeration economies are fully to exert their beneficial effects. The quality of the activities hosted, the quality of production factors, the density of external linkages and cooperation networks, the quality of urban infrastructure—for internal and external mobility, education, public services—are all factors enabling a long-term "structural dynamics" process (in the language of dynamic modelling) via what can be called a process of urban evolution and transformation.

In this sense, the explanation of a relatively good urban economic performance is not mechanically linked to the existence of agglomeration economies. Instead, this approach highlights the conditions under which agglomeration economies manifest themselves and may be fully exploited within each urban size class.

This approach confirms the existence of agglomeration economies, as well as the risk of agglomeration diseconomies, but this *general law works within each class of cities*. There are large cities able to escape agglomeration diseconomies, despite their huge size, and there are small cities that record decreasing returns in spite of their small size. The explanation of this apparent contradiction lies in the capacities of cities to overcome diseconomies of scale either by innovating in the functions they host or by launching cooperation through networks with other cities. In other words, the explanation lies in the presence of strategic territorial capital assets.

The concept of city networks, already present in the literature for many years (Camagni, 1993), recurs in the explanation of urban dynamics. The concept of "city networks" adds to that of "borrowed size" the idea that size can be borrowed not only thanks to physical proximity to larger centres, but also thanks to relationships and flows of a mainly horizontal and non-hierarchical nature among complementary or similar centres, located far from each other, with the purpose of achieving network externalities (Camagni, 1993; Camagni & Capello, 2004; Capello, 2000): a process that could be termed "borrowed functions".

As Camagni *et al.* claim in this issue, these conceptual ideas help explain why cities of intermediate size are being increasingly looked upon as the places that could well host the growth of the years to come: limited city size, in fact, facilitates environmental equilibrium, efficiency of the mobility system and the possibility for citizens to maintain a sense of identity, provided that a superior economic efficiency is reached ("borrowed") through external cooperation with other cities located in the same region, or are distant but well connected with it.

In sum, this approach highlights that there exists a *universal law of agglomeration economies that applies across all cities of any size, showing marked specificities within each size class. Within each city class, the quality of territorial capital assets—presence of high-value functions or networking and cooperation capabilities—is the condition sine qua non to avoid entering a phase of decreasing returns.* In this perspective, smaller cities have high potential for growth if they are able to enter a virtuous and cumulative path of transformation and innovation through the exploitation of high-quality territorial assets in spite of their limited size.

6. From Size to Growth Through Efficiency: A Comparison Among the Three Approaches

The existence of these three approaches calls for a final comparison of their capacities to interpret urban dynamics through a different conceptualization of agglomeration economies.

In the logic of the micro-economic approach, the size of a city (and of industries within the city) determines its static efficiency; greater efficiency (productivity) generates firms' competitiveness, which in its turn explains growth opportunities for the place in which firms are located (Table 1). On this reasoning, pecuniary externalities play a major role in explaining agglomeration forces, restricting technological externalities to a specific and limited case: that of knowledge spillovers as sources of agglomeration economies (van Oort *et al.*, 2014). The result is that "agglomeration advantage" refers only to a mere input-output relationship among clustered firms. The urban context does not influence firms' performance, and it benefits only indirectly from a quantitative expansion. Cities are equated to agglomerations of firms; territory is downscaled to physical distance or geometric space.

In the geographical approach, interpretation of urban dynamics through agglomeration economies is based on a direct logical link: size and proximity generate technological externalities that explain urban competitiveness and therefore growth (Table 1). What allows the direct link to occur is the introduction of geographical space (geographical and cognitive proximity, and not solely the size of the urban production complex), as a source of externality and consequently the driving force behind the physical expansion of cities. However, the approach confines urban dynamics to the pure physical evolution of cities and urban systems, with no role assigned to their possible structural evolution and transformation.

Instead, internal structural transformation is central in the third approach, the macro-territorial approach. Thanks to its interpretation of the role of territorial capital in the capacity to exploit increasing returns to urban size, this approach is able to explain the structural dynamics of cities (and of urban systems) (Table 1). The capacity of cities to overcome diseconomies of scale is linked to the presence of strategic territorial capital assets allowing either innovation in the functions hosted or achievement of network externalities through cooperation with other cities. These elements directly affect urban competitiveness and growth, generating what in mathematical ecology is called "structural dynamics".

The differences underlined stress the strong complementarities among the three approaches; they all increase the interpretive capacity of agglomeration theories in the explanation of urban dynamics. But a *direct* link between agglomeration and urban dynamics is obtained only if technological externalities (mainly due to proximity and synergies) are considered. However, if agglomeration economies are to interpret structural evolutions and transformation, the role of the territory—defined as a set of cognitive, institutional, physical and natural local assets potentially generating competitiveness and utility flows to local communities—has to be taken into account.

7. Initial Policy Implications

An initial interesting and innovative policy lesson emerges from the present special issue. Betting and investing on second-rank cities seems to be a sensible strategy, provided that an evolutionary and innovation-oriented perspective is adopted. Particularly in emerging countries (like the EU Central Eastern European Countries), a national strategy based on the provision of basic accessibility and human capital infrastructure may prove more forward-looking with respect to investing only in a few capital cities. In fact, it may prevent the drawbacks inevitably linked to an excessive concentration of development resources in a few sports, namely wage and price increases, congestion and all diseconomies related to size and "sudden" urban growth (Camagni, 2008; Camagni & Capello, 2013).

Investing in second-rank cities allows a wider and faster exploitation of dispersed and not fully exploited territorial capital assets, thanks to the capability of identifying and engaging these assets by local stakeholders and dynamic local élites, especially in times of consolidation and diffusion of technological paradigms, like the present knowledge economy one. This statement finds support and empirical corroboration in some very recent econometric research work on regional development and scenario foresights in the EU: a scenario encompassing consistent policies addressed at strengthening second-rank city-regions turns out to be more expansionary (and also more cohesive) than a scenario of concentration of public investments in capital cities and large, advanced city-regions (Camagni & Capello, 2014). Similar policy messages come from many papers collected in this special issue (Agnoletti *et al.*, 2014; Brezzi & Veneri, 2014; Parkinson *et al.*, 2014).

8. Aims and Structure of the Special Issue

This special issue goes deeply into the literature reviewed above by offering a collection of original papers pursuing two main inter-related aims. The first is to identify the reasons for the dynamics of second-rank cities.[3] In particular, in light of the theoretical discussions presented in the previous sections, this issue contains papers that explore the geographical and macro-territorial foundations of agglomeration economies. The second aim is to highlight the role of second-tier cities in polycentric urban structures, capturing how urban externality fields and network externalities play a role in polycentric areas with different hierarchical structures.

An introductory paper by Michael Parkinson, Richard Meegan and Jay Karecha provides data on the relatively good performance of second-rank cities (in their definition, non-capital cities) in Europe, and asks a simple but crucial question: why should governments invest outside their capitals? This is a question that assumes significant weight in a period of extremely scarce public resources like the one characterized by the current economic downturn.

The three papers that follow initiate analyses of how agglomeration economies explain second-rank urban dynamics. The paper by Roberto Camagni, Roberta Capello and Andrea Caragliu presents a macro-territorial conceptualization, claiming that second-rank cities may enter a phase of decreasing returns despite their size. The explanation of this apparent paradox is that cities are unable to improve territorial capital assets. They do not innovate in the functions that they perform, or in the organization of activities with other cities, when they are unable to offer these functions on their own. The main results of the empirical evidence reported on 136 European cities is that outperforming second-rank cities are those characterized by economies of scale, and that these economies of scale are related to territorial capital assets, like the level of functions and networks possessed by cities.

Martijn Burger, Evert Meijers, Marloes Hoogerbrugge and Jaume Masip Tresserra analyse the geographical foundations of agglomeration economies by treating the concept of "borrowed size" as essential for understanding urban patterns and dynamics in North-West Europe. More importantly, they make a first attempt to measure borrowed size vs. agglomeration shadows by resorting to the sizes of cities and their positions in the urban hierarchy as explanations for a city's capacity to borrow size. The main findings are in fact that the largest cities in a functional urban area obtain the highest benefits from the size of the rest of the area in which they are located and from access to a large (inter)national market. Lower order places are less likely to borrow size from other places. They remain in the shadow of a first-rank city and undergo spatial competition effects.

The paper by Frank van Oort, Stefan de Geus and Teodora Dogaru focuses on the agglomeration circumstances influencing economic growth across European urban regions. The novelty of the paper is that it suggests a possible way out of the current seemingly locked-in debate on the relation between agglomeration and growth, which is ambiguous and indecisive with regard to whether specialization or diversity is facilitated by (sheer) urbanization. The debate on whether specialization or diversity explains urban dynamics is resolved by the introduction of concepts like related and unrelated variety in the empirical modelling of growth across European regions, arguing that it is not simply the presence of different technological or industrial sectors that will yield positive results; rather, sectors require complementarities that exist in terms of shared competences (Frenken et al., 2007).

In the second part of the special issue, attention is concentrated on the role of second-rank cities in polycentric urban structures at different (national and regional) levels. Monica Brezzi and Paolo Veneri present an empirical analysis on the role of national and regional polycentricity in explaining the economic well-being of regions and nations in all OECD countries. Interestingly, national polycentricity (i.e. a network of cities) is found to be positively correlated with high per capita GDP levels. Instead, at the regional level, polycentric urban structures seem to have lower per capita GDP levels than monocentric ones: pure agglomeration economies seem to generate higher economic efficiency and greater well-being than borrowed size.

Chiara Agnoletti, Chiara Bocci, Sabrina Iommi, Patrizia Lattarulo and Donatella Marinari conduct an interesting discussion on economic efficiency (in terms of the competitiveness of urban services produced) and sustainability (in terms of land consumption and urban fragmentation) of different hierarchical models within polycentric urban structures. Drawing on a rich dataset on Italian cities, the empirical analysis shows that regions with a polycentric structure consisting of small- and medium-sized cities are rather competitive and sustainable. This leads to the rather new and interesting result that the size of cities alone does not explain differences in the economic and environmental performances of polycentric urban systems.

Last, but not least, Rudolf Giffinger and Johannes Suitner debate whether metropolization should be analysed in terms of a process rather than a state. This approach is essential to grasp the structural differences in the metropolization process in Europe, and to highlight the fact that European city-regions have reached different stages of polycentric metropolitan development. This perspective is also an essential foundation for learning processes in the governance of future polycentric metropolitan development.

Notes

1. See among others Ciccone (2002), Ciccone and Hall (1996), Ellison and Glaeser(1997), Henderson (2003), Rosenthal and Strange (2001, 2003).
2. Some scholars have already suggested that agglomeration costs are more confined to city boundaries than agglomeration benefits (Parr, 2002).
3. The papers contained in this special issues were presented in draft versions at the international seminar on "Welfare and competitiveness in the European polycentric urban structure", organized by IRPET, held in Florence on 7 June 2013. The editors are grateful to Patrizia Lattarulo of IRPET for her scientific and financial help in the organization of the international seminar.

References

Agnoletti, C., Bocci, C., Iommi, S., Lattarulo, P. & Marinari, D. (2014) First and second tier cities in regional agglomeration models, *European Planning Studies*. doi:10.1080/09654313.2014.905006

Alonso, W. (1973) Urban zero population growth, *Daedalus*, 102(4), pp. 191–206.

Azzoni, C. (1986) Indústria e reversião de polarizaciõn de Brasil, Essays in Economics no. 58, Institute of Economic Studies, University of Saõ Paulo, Saõ Paulo.

Beaudry, C. & Schiffauerova, A. (2009) Who's right, Marshall or Jacobs? The localization versus urbanization debate, *Research Policy*, 38(2), pp. 318–337.

Becattini, G. (1989) Sectors and/or districts: Some remarks on the conceptual foundations of industrial economics, in: E. Goodman & J. Bamford (Eds) *Small Firms and Industrial Districts in Italy*, pp. 123–135 (London: Routledge).

Brezzi, M. & Veneri, P. (2014) Assessing polycentric urban systems in the OECD: Country, regional and metropolitan perspectives, *European Planning Studies*. doi:10.1080/09654313.2014.905005

Burger, M. J., Meijers, E. J., Hoogerbrugge, M. M. & Tresserra, J. M. (2014) Borrowed size, agglomeration shadows and cultural amenities in Western Europe, *European Planning Studies*. doi:10.1080/09654313.2014.905002

Camagni, R. (1991) Technological change, uncertainty and innovation networks: Towards a dynamic theory of economic space, in: Camagni, R. (Ed) *Innovation Networks: Spatial Perspectives*, pp. 121–144 (London: Belhaven-Pinter).

Camagni, R. (1993) From city hierarchy to city networks: Reflection about an emerging paradigm, in: T. Lakshmanan & P. Nijkamp (Eds) *Structure and Change in the Space Economy: festschrifts in Honour of Martin Beckmann*, pp. 66–87 (Berlin, DE: Springer-Verlag).

Camagni, R. (2008) Towards a conclusion: Regional and territorial policy recommendations, in: R. Capello, R. Camagni, U. Fratesi & B. Chizzolini (Eds) *Modelling Regional Scenarios for the Enlarged Europe*, pp. 283–306 (Berlin: Springer-Verlag).

Camagni, R. & Capello, R. (2004) The city network paradigm: Theory and empirical evidence, in: R. Capello, P. Nijkamp (Eds) *Urban Dynamics and Growth: Advances in Urban Economics*, pp. 495–532 (Amsterdam: Elsevier).

Camagni, R. & Capello, R. (2013) Regional innovation patterns and the EU regional policy reform: Towards smart innovation policies, *Growth and Change*, 44(2), pp. 355–389.

Camagni, R. & Capello, R. (2014) Rationale and Design of EU Cohesion Policies in a Period of Crisis. Paper presented at the International Seminar of GRINCOH, Milan, February 27–28.

Camagni, R., Capello, R. & Caragliu, A. (2013) One or infinite optimal city sizes? In search of an equilibrium size for cities, *The Annals of Regional Science*, 51(2), pp. 309–341.

Camagni, R., Capello, R. & Caragliu, A. (2014a) Structural Dynamics of Second vs. First Rank Cities: Similar Laws, High Specificities. Paper presented at the 60 NARSC conference, Atlanta, GA, November 13–16.

Camagni, R., Capello, R. & Caragliu, A. (2014b) The rise of second-rank cities: What role for agglomeration economies? *European Planning Studies*. doi:10.1080/09654313.2014.904999

Capello, R. (2000) The city network paradigm: Measuring urban network externalities, *Urban Studies*, 37(11), pp. 1925–1945.

Capello, R. (2009) Indivisibilities, synergy and proximity: The need for an integrated approach to agglomeration economies, *Tijdschrift voor Economische en Sociale Geographie (TESG)*, 100(2), pp. 145–159.

Carlino, G. (1980) Contrasts in agglomeration: New York and pittsburgh reconsidered, *Urban Studies*, 17(3), pp. 343–351.

Christaller, W. (1933) *Die Zentralen Orte in Süddeutschland [The Central Places in Southern Germany]*, Wissenschaftliche Buchgesellschaft, Darmstadt, English edition (1966), Prentice-Hall, Englewood Cliffs, NJ.

Ciccone, A. (2002) Agglomeration effects in Europe, *European Economic Review*, 46(2), pp. 213–227.

Ciccone, A. & Hall, R. E. (1996) Productivity and the density of economic activity, *American Economic Review*, 86(1), pp. 54–70.

Cohen, J. & Paul Morrison, C. (2009) Agglomeration, productivity and regional growth: Production theory approaches, in: R. Capello & P. Nijkamp (Eds) *Handbook of Regional Dynamics and Growth: Advances in Regional Economics*, pp. 101–117 (Cheltenham: Edward Elgar).

Conti, S. & Dematteis, G. (1995) Enterprises, systems and network dynamics: The challenge of complexity, in: S. Conti, E. Malecki & P. Oinas (Eds) *The Industrial Enterprise and Its Environment: Spatial Perspectives*, pp. 217–242 (Aldershot: Avebury).

Dijkstra, L., Garcilazo, E. & McCann, P. (2013) The economic performance of European cities and city regions: Myths and realities, *European Planning Studies*, 21(3), pp. 334–354.

Dobkins, L. H. & Ioannides, Y. M. (2001) Spatial interaction among U.S. cities: 1900–1990, *Regional Science and Urban Economics*, 31(6), pp. 701–731.

Ellison, G. & Glaeser, E. L. (1997) Geographic concentration in US manufacturing industries: A dartboard approach, *Journal of Political Economy*, 105(5), pp. 889–927.

Faggio, G., Silva, O. & Strange, W. (2013) Heterogeneous Industries. Paper presented at the 60th North American Regional Science Conference, Atlanta, GA, November 13–16.

Frenken, K., van Oort, F. G. & Verburg, T. (2007) Related variety, unrelated variety and regional economic growth, *Regional Studies*, 41(5), pp. 685–697.

Friedmann, J. (1986) The world city hypothesis, *Development and Change*, 17(1), pp. 69–83.

Fujita, M. & Mori, T. (1996) The role of ports in the making of major cities: Self-organization and hub effects, *Journal of Development Economics*, 49(1), pp. 93–120.

Fujita, M. & Mori, T. (1997) Structural stability and evolution of urban systems, *Regional Science and Urban Economics*, 27(4), pp. 399–442.

Giffinger, R. & Suitner, J. (2014) Polycentric metropolitan development: From structural assessment to processual dimension, *European Planning Studies*. doi:10.1080/09654313.2014.905007

Henderson, J. (1974) The sizes and types of cities, *The American Economic Review*, 64(4), pp. 640–656.

Henderson, J. (1985) *Economic Theory and the Cities* (Orlando, FL: Academic Press).

Henderson, J. V. (2003) Marshall's scale economies, *Journal of Urban Economics*, 53(1), pp. 1–28.

Hoch, I. (1972) Income and city size, *Urban Studies*, 9(3), pp. 299–328.

Isard, W. & Schooler, E. (1959) Industrial complex analysis, agglomeration economies and regional development, *Journal of Regional Science*, 1(2), pp. 19–33.

Krugman, P. (1991) *Geography and Trade*, (Cambridge, MA: MIT Press).

Krugman, P. (1993) First nature, second nature and metroplitan location, *Journal of Regional Science*, 33(2), pp. 129–144.

Markusen, A., Lee, Y.-S. & Di Giovanna, S. (Eds) (1999) *Second-Tier Cities: Rapid Growth Beyond the Metropolis*, (Minneapolis, MN: University of Minnesota Press).

Meijers, E. (2013) Cities Borrowing Size: An Exploration of the Spread of Metropolitan Amenities across European Cities. Paper presented at the Association of American Geographers Annual Meeting, Los Angeles, April 9–13.

Mera, K. (1973) On the urban agglomeration and economic efficiency, *Economic Development and Cultural Change*, 21(2), pp. 309–324.

Mills, E. (1970) Urban density functions, *Urban Studies*, 7(1), pp. 5–20.

Moomaw, R. (1983) Is population scale worthless surrogate for business agglomeration economies? *Regional Science and Urban Economics*, 13(4), pp. 525–545.

Parkinson, M., Meegan, R. & Karecha, J. (2014) City size and economic performance: Is bigger better, small more beautiful or middling marvellous? *European Planning Studies*. doi:10.1080/09654313.2014.904998

Parr, J. B. (2002) Agglomeration economies: Ambiguities and confusions, *Environment and Planning A*, 34(4), pp. 717–731.

Partridge, M. (2010) The dueling models: NEG vs amenity migration in explaining U.S. engines of growth, *Papers in Regional Science*, 89(3), pp. 513–536.

Partridge, M., Rickman, D. S., Ali, K. & Olfert, M. R. (2009) Do new economic geography agglomeration shadows underlie current population dynamics across the urban hierarchy? *Papers in Regional Science*, 88(2), pp. 445–467.

Phelps, N. A., Fallon, R. J. & Williams, C. L. (2001) Small firms, borrowed size and the urban-rural shift, *Regional Studies*, 35(7), pp. 613–624.

Richardson, H. W. (1980) Polarization reversal in developing countries, *Papers of the Regional Science Association*, 45(1), pp. 67–85.

Rosenthal, S. S. & Strange, W. C. (2001) The determinants of agglomeration, *Journal of Urban Economics*, 50(2), pp. 191–229.

Rosenthal, S. S. & Strange, W. C. (2003) Geography, industrial organization, and agglomeration, *Review of Economics and Statistics*, 85(2), pp. 377–393.

Sassen, S. (1991) *The Global City* (Princeton, NJ: Princeton University Press).

Scott, A. (2001) *Global City-Regions* (Oxford: Oxford University Press).

Segal, D. (1976) Are there returns to scale in city size? *Review of Economics and Statistics*, 58(3), pp. 339–350.

Shefer, D. (1973) Localization economies in SMSA'S: A production function analysis, *Journal of Regional Science*, 13(1), pp. 55–64.

Storper, M. (1995) La Géographie des Conventions: Proximité Territoriale, Interdépendences Non-Marchandes et Développement Economique, in: A. Rallet & A. Torre (Eds) *Economie Industrielle et Economie Spatiale*, pp. 111–128. (Paris: Economica).

Sweikauskas, L. (1975) The productivity of city size, *Quarterly Journal of Economics*, 89(3), pp. 393–413.

Townroe, P. & Kean, D. (1984) Polarization reversal in the state of sao paulo, *Brazil Regional Studies*, 18(1), pp. 45–54.

Van Oort, F., de Geus, S. & Dogaru, T. (2014) Related variety, regional economic growth and place-based development—strategies in urban networks in Europe, *European Planning Studies*. doi:10.1080/09654313.2014.905003

City Size and Economic Performance: Is Bigger Better, Small More Beautiful or Middling Marvellous?

MICHAEL PARKINSON, RICHARD MEEGAN & JAY KARECHA

European Institute for Urban Affairs, Liverpool John Moores University, Liverpool, UK

ABSTRACT   *This article discusses the contribution that second-tier cities can and do make to the economic performance of national economies across Europe. It reviews the competing theories about size, investment and economic performance. It presents a range of evidence about the performance of over 150 European capital and second-tier cities in 31 countries. It identifies some key policy messages for local national and European policy-makers. It presents evidence that decentralizing responsibilities, powers and resources, spreading investment and encouraging high performance in a range of cities rather than concentrating on the capital city produces national benefits. It argues that in a period of austerity national governments should resist pressures to concentrate investment in capital cities and invest more in second-tier cities when there is evidence that: (i) the gap with capitals is large and growing (ii) the business infrastructure of second-tier cities is weak because of national underinvestment and (iii) there is clear evidence about the negative externalities of capital city growth. It argues that the issues have slipped down the European Commission's agenda and it should do more to ensure its strategies help realize the economic potential of second-tier cities in future.*

1. The Recession Makes the Discussion of City Size More Important

This special edition focuses upon the economic contribution and performance of different-sized cities across Europe. The issues have always been academically significant, but the current recession makes them even more significant politically. They have sharpened the debate about policies for national competitiveness. They have also sharpened the debate about the economic contribution of capital and second-tier cities and whether countries perform better if they concentrate their investment in their national capitals or spread investment across a wider set of cities. Does size matter—and how much—for

14

cities and national economic performance? They pose a single crucial question: "Why should policy-makers invest beyond the capital cities in an age of austerity?" In the recent past a major focus of academic research has been on the larger global cites (Sassen, 2001, 2012; Brenner & Keil, 2006). However, the new economic conditions mean policy-makers will have to pay more attention in future to the potential and actual contribution of cities further down the urban hierarchy—the second-tier cities.

The global recession and Eurozone crisis have had a huge impact upon the European economy and present great future threats. This debate will become more important during the next decade as the crisis threatens to undermine the real achievements made by many European cities. In the past decade, cities in many countries improved their economic performance and made a growing contribution to national competitiveness. But it was a result of high-performing national economies and substantial investment of public resources. Those conditions will not be found during the next decade. Many underlying economic and social problems in cities—which had been masked by the boom—have already been intensified by the crisis. There is a risk that economic and fiscal problems and the competition for scarce public and private sector resources will limit the growth of cities and widen economic and social gaps within them and between them and the capitals. The investment that was made during the growth years before the crash paid off, improving the economic performance and national contribution of many second-tier cities. It is crucial that the investment that was made—and which is now at risk in the recession—is not lost. The argument is essentially one about investment and economic performance as much as territorial justice.

This article explores some of the policy and research questions raised by this debate, based upon a major study we recently conducted of European cities in the age of austerity.

2. What Are the Analytical Arguments About City Size, Investment and Economic Performance—Place-Neutral vs. "Place-Based" Policies?

2.1. Competing Theories

There has been a rediscovery of the importance of agglomeration and urbanization economies and externalities in urban and regional economic growth. A number of different theoretical frameworks compete for attention including, notably, neo-classically based endogenous growth theory, geographical economics and institutional and evolutionary theories. Geographical economics and the so-called "New Economic Geography" focus, for example, on the external economies and increasing returns to scale associated with regional industrial specialization and concentration and the urbanization economies from agglomeration of firms from different industries that underpin the growth of urban locations (Krugman, 1990, 1991, 1993; Fujita et al., 1999; Duranton & Puga, 2004; Kitson et al., 2004; World Bank, 2009). Agglomeration economies also feature in macro-structural economic transition theories which link local and regional growth potential to the transition from the macro-economic era of mass production to the current era of "flexible specialisation" (Piore & Sabel, 1984; Storper & Scott, 1988; Scott, 1988). Institutional and evolutionary theories of regional economic development have focused on the institutional arrangements and "softer" factors like networking, trust and social capital that together provide externalities that encourage the emergence and subsequent growth of local and regional economies (Grabher, 1993; Amin & Thrift, 1995; Maskell, 2002).

15

2.2. *Competing Policy Paradigms*

These theoretical differences translate into different policy recommendations. Free market neo-classical economists stress the importance of agglomeration economies as the justification for allowing capital cities to grow in an unrestricted fashion to reflect market demand and forces. In this view, capital cities have significant agglomeration advantages. They are typically the centres of national political, administrative and economic power. They have stronger private sectors. They are more integrated into global networks. They are more likely to contain companies' headquarters. Their producer services are typically the most advanced. They contain major financial institutions which provide easier access to risk capital. They contain leading academic and research institutions. They are at the hub of national transportation and ICT networks. They attract public and private "prestige" investment because they "represent" their nations. Henderson, for example, argues that capital cities receive preferential treatment from national governments because public decision-makers find it easier to allocate resources to existing capitals rather than identify opportunities elsewhere (Henderson, 2010). Similarly, it has been argued that private sector investors adopt the safer strategy of investing in buoyant, capital locations rather than taking risks with more distant, perhaps more economically marginal locations.

The neo-classical position also encourages a "place-neutral" approach to economic development best illustrated in the World Bank's arguments that that policies should emphasize people over place and that, because growth and development are inevitably unbalanced, it is counterproductive to attempt to shift that market balance (World Bank, 2009). Urban and regional development policy should emphasize "space-blind" provision of universal public services like education and social services and general infrastructure investment with only very limited use of explicitly spatially targeted interventions (Hildreth & Bailey, 2013).

The "place-neutral" approach is challenged from an institutional and evolutionary economic geography perspective, which emphasizes the costs and negative externalities of agglomeration. Agglomeration clearly produces economic benefits. However, the economic benefits of agglomeration are not unlimited. Capital cities can reach a point where diseconomies make them less competitive because of the negative externalities caused by unregulated growth and diminishing marginal returns. The OECD has made a significant contribution to this debate with a series of studies exploring the contribution of different regions to national competitiveness. Some of its recent work has focussed specifically upon the middle regions, showing that growth does not come only from a small number of leading regions at the top but from the many more regions further down a long territorial tail of the regional hierarchy, whose collective contribution is crucial. OECD's policy position is that the economic contribution of the middle regions is typically underestimated and governments should do more to maximize their contribution if they want to maximize national competitiveness (OECD, 2006, 2012a, 2012b; Garcilazo et al., 2010).

From this perspective, urban and regional policies need to be more sensitive to local context and local specificity, "place-based" as opposed to "place neutral" (Barca, 2009; Barca & McCann, 2010; McCann & Ortega-Argilés, 2011; McCann & Rodríguez-Pose, 2011; Barca et al., 2012). "Bottom-up" policies that take into consideration, for example, the localized forces that influence innovation and development are needed, albeit reconciled with "top-down" policies in an approach to urban and regional development issues from a meso-level perspective (Crescenzi & Rodríguez-Pose, 2011). The emphasis on "place-based" policy also explains the growing demand for decentralization of powers from national to sub-national governments (Ascani et al., 2012). Our work supports this latter view.

3. What Was the Economic Contribution of Capital and Second-Tier Cities in the European Boom and Recession?

This article draws on a recent study for ESPON of 124 second-tier and 31 capital cities across Europe (Parkinson et al., 2012; Parkinson & Meegan, 2013). There are many typologies of cities. All have their limitations. We use the concept of second-tier cities—those cities outside the capital whose economic and social performance is sufficiently important to affect the potential performance of the national economy. It does not imply that they are less important than the capital cities. It certainly does not mean that they are second class. And it does not mean they are the "second" city—because there is only one of these in each country. And second-tier cities are not all the same—they vary enormously. Sometimes they are very large regional capitals. Sometimes they are the second largest city of the country with huge national significance—for example Barcelona, Munich and Lyon. But many are much smaller. However, while they differ in many respects, second-tier cities can play comparable national economic roles. The second-tier cities we studied in the research for ESPON constitute almost 80% of Europe's metropolitan urban population. They lie between the capital cities which contribute a huge amount to their national economy and the many smaller places which contribute rather less. They are the middle of the urban system.

We recognize that the distinction between capital and second-tier cities is an administrative rather than a functional one. However, we use it at least partly because a key policy concern of our work is the extent to which national governments focus attention and resources on their capitals rather than on other cities in their urban system—and the consequences for national economic performance. There is some argument that they do so (Henderson, 2010). Also this is a hugely significant policy issue for the European Commission who commissioned the research. They were aware that many member states, especially in the east of Europe, favoured investing EU resources in their capital cities on the grounds it would lead to greater national economic gains. We wanted to test that argument. However, although partly an administrative distinction, in the vast bulk of countries the capital is by far the largest city economically and demographically. So we are effectively examining relationships between economically dominant cities and their urban systems in European countries. We recognize this is not the first time this concept has been used. Markusen et al. (1999) looked at the phenomenon of second-tier cities. However, their work primarily focussed on non-European cities and they were typically more recently developed and often smaller cities than ours. This work adds new information and develops the concept.

3.1. Decentralization and Deconcentration Matter

What are our key messages? Our evidence shows that decentralizing responsibilities, powers and resources, spreading investment and encouraging high performance in a range of cities rather than concentrating on the capital city produces national benefits. In terms of policy, some countries concentrate attention and resources on the capitals at the expense of their second-tier cities. But many are beginning to develop policies which explicitly target second tiers. More widely, in some countries mainstream national policies which implicitly affect urban competitiveness—innovation, diversity, skills, connectivity, place quality and strategic governance capacity—have been used to help second-tier cities develop. Most interestingly, in countries which are less centralized and less economically concentrated, and where cities have greater powers, resources and responsibilities, cities have performed better and helped the national economy more. There is evidence that levels of government decentralization do matter. Between 2000 and 2007 for example, in the Federal states, all German and Austrian and half of Belgium's second-tier cities outperformed their capitals. In the regionalized states, all Spanish and a third of Italian second-tier cities grew faster than their capitals. In the Nordic states, all grew faster than their capital. In the unitary centralized states of Hungary, Hungary, Slovakia, Slovenia, Estonia, Lithuania and Bulgaria all second-tier cities and all but one in the Czech Republic had lower growth rates than their capital cities. Only in Romania, Latvia and Croatia did some second-tier cities outperform their capital.

Our study and this article argue that continuing over-investment in capital cities and under-investment in second-tier cities in the long run will be unsustainable and lead to economic under-performance. It finds much evidence that decentralizing responsibilities, powers and resources, spreading investment and encouraging high performance in a range of cities rather than concentrating on the capital city produces national benefits. Although the capital cities in many countries are responsible for a significant proportion of national GDP, second-tier cities still make a large contribution. In many cases the collective economic contribution that second-tier cities make is greater than that of the capital itself. Individually, second-tier cities may lag behind capitals. But collectively their contribution to national economic performance is hugely significant.

However, we do not claim, and our evidence does not show, that every second-tier city in every country performed well in the boom or recession or that they outperform the capital city. But enough cities in enough countries have performed well enough to challenge the assumption that capital cities should be the first choice for investment to achieve national economic success. Our work does present a huge amount of compelling evidence from quantitative data analysis, policy reviews and individual city studies that point in the same direction. We can only present a limited amount of that quantitative evidence in this article. They cumulatively demonstrate that policy-makers should take these issues more seriously in future and systematically examine how their decisions affect second-tier cities.

3.2. *Germany Proves the Point*

Germany provides important lessons on the economic role of second-tier cities. Of course Germany is unique. It is a Federal system which has changed its capital. The country has been divided and reunited. Its second-tier cities are typically state capitals with extensive powers and resources. It has a unique system of regional banking and powerful middle-sized firms. It is not possible for other European countries to simply imitate the structural characteristics of the German system. Nevertheless, the key principles of the German experience can be transferred between different countries. Its experience particularly underlines the argument that decentralization of powers and resources and the spatial deconcentration of investment leads to a higher performing national economy. Economic activity—private and public—is more evenly distributed across a range of cities that form a powerful multi-cylinder, economic engine. Between 2000 and 2007 populations increased faster in six German second-tier cities than in Berlin. All its 14 second-tier cities had productivity growth rates above those in Berlin. At a European level, 5 of the top 10 second tiers in GDP growth between 2000 and 2007 were German. Five of the top ten most innovative cities were German.

3.3. *Capital Cities Dominate but Second-Tier Cities Make an Important Contribution to Competitiveness*

The essential picture is that—with the crucial exception of Germany—capital cities dominate the European urban system in terms of population, employment and output. The gap between capital and second-tier cities is large and in virtually all the former socialist states of Eastern Europe, it is growing. The total GDP of capital cities in 2007 was greater than their leading second-tier cities in all but two countries, Germany and Italy (Figure 1). In 19 countries the total GDP of the capital was more than twice that of the largest non-capital city and was as much as 8 times greater in 4 states—UK, France, Hungary and Latvia. Nevertheless our evidence shows that all second-tier cities made a contribution—and some a significant one—to economic growth in Europe between 2000 and 2007, even if many were overshadowed by capital cities to different degrees in different parts of Europe. The size of the gap between capitals and secondaries varies and in some cases is declining.

Structurally, capitals dominate their national economies. But change is also important. And many second-tier cities improved their position in the boom years 2000–2007. In 16 of the 26 countries, 1 or more second-tier cities had annual GDP growth higher than their capitals. In Austria and Germany, all second-tier cities outperformed their capitals. The relatively strong growth rates in a number of capitals and second-tier cities in Eastern Europe, as their economies integrated into the European economy, also stand out. Indeed, the highest growth rates over this period were found there (Figure 2).

3.4. *Second-Tier City Growth During the Boom—Regionally Differentiated*

Although many second-tier cities performed well during the boom years when they had national government support and investment, there were large regional disparities (Map 1). The prosperous city-regions in the North, Central and West of Europe contrast with the less prosperous city-regions of the Central East, East and South East. Together, the North, Central and West groupings housed over four fifths of the leading city-regions. By contrast only 3 of the 27 city-regions in the former socialist Unitary states of Czech Republic, Hungary, Poland, Slovakia and Slovenia were leading, 3 were intermediate and the great majority were lagging. There were some shifts during the period. The lagging city-regions of East and South East Europe, and to a lesser extent, of Central East Europe had a rapid burst of growth in the eight-year period. Those in

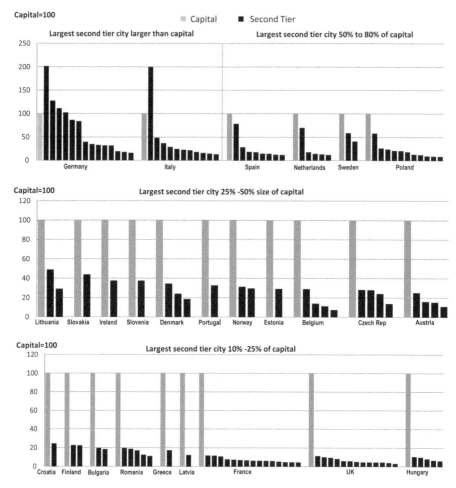

Figure 1. Total GDP in PPS, 2007.
Source: Parkinson et al. (2012).

transition to capitalist economies and integrating into the European economy had average growth rates 6 and 7 times those of the established West and North.

3.5. *The Crisis Threatens to Undermine Achievements of Second-Tier Cities*

But the recession has had a major impact on many of them—in particular those which flourished during the boom decade (Map 2). More than 75% of the cities experienced GDP falls 2008–2010. Capitals performed far better than second-tier cities during the crisis. The better-performing places were in Eastern Europe and in Poland in particular. The fastest growing 19 places—12 Polish—were all in Eastern Europe. The Baltics have been heavily hit. Major Western European countries have all been hit. In Germany only Berlin grew. In all other German cities GDP declined. In the UK all 14 cities declined. In Italy all 12 cities declined. In Spain 8 of 9 declined.

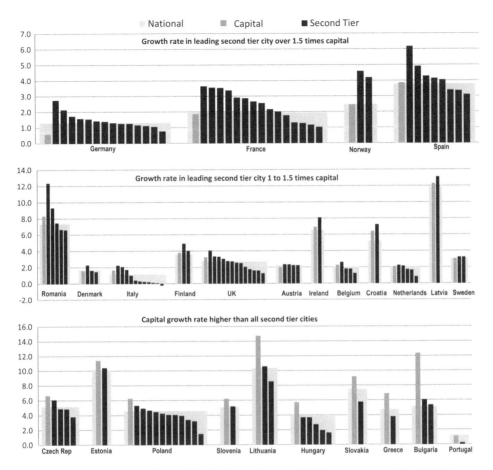

Figure 2. Second-tier cities' contribution increasing: total GDP average annual real % change 2000–2007.
Source: Eurostat.

4. So What Are the Policy Messages for Governments About Second-Tier Cities?

4.1. *Decentralization of Powers and Resources—to Match Devolved Responsibilities— Is Crucial*

Levels of centralization matter. But decentralization of responsibilities to cities only works if responsibilities are matched by corresponding powers and resources. Cities perform better in those countries which are less centralized and economically concentrated and where cities have greater powers, resources and responsibilities. Many policy-makers and researchers believe that, given the impact of deconcentrating resources and decentralizing powers on second-tier cities, national policies should give them more powers, responsibilities and resources. A policy of economic place making has benefitted many cities. It underlines the potential for wider implementation in national and European policies in the future.

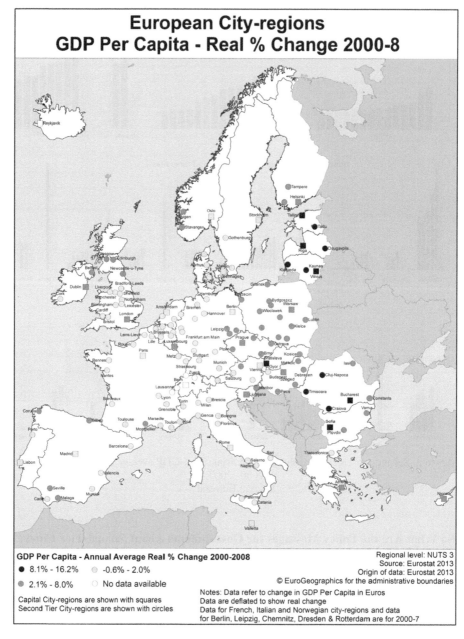

Map 1. European city-regions' economic growth, 2000–2008.
Source: Parkinson et al. (2014).

4.2. *The Limits of Capital Cities*

Our study also identified a series of concerns about the dominance of capital cities. One theme is the costs and negative externalities of agglomeration. The second theme is that all urban areas have potential which national policy should encourage, rather

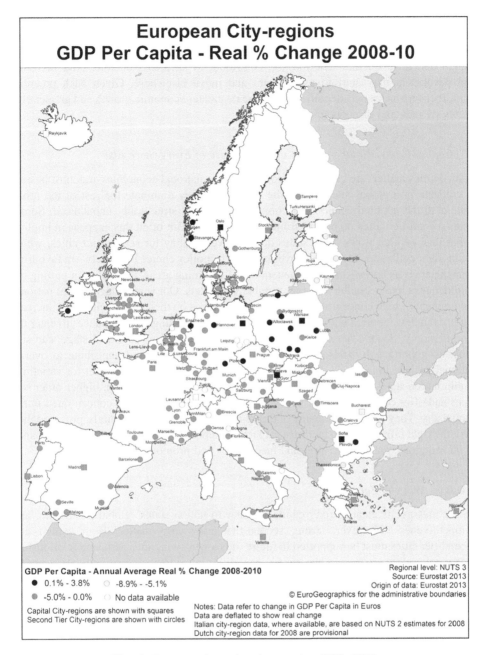

Map 2. European city-regions in recession, 2008–2010.
Source: Parkinson et al. (2014).

than concentrating upon a limited number of already successful places. Agglomeration does obviously produce economic benefits. OECD research has shown that in some countries, the largest single metropolitan area produces between one-third and one-half of national GDP. However, the economic benefits of agglomeration are not limitless.

Cities can reach a point where external diseconomies make them less competitive because of negative externalities caused by unregulated urban growth and diminishing marginal returns. Beyond a certain point, congestion, land scarcity, sprawl, marginalized human capital and infrastructure deterioration contribute to an area's decline. And investors and developers may start to avoid them and move elsewhere. Given such potential risks, focusing on second-tier cities would create greater economic growth and greater efficiency by reducing diseconomies of scale.

4.3. *Capital Cities Matter—but Not at the Expense of Everywhere Else*

Capital cities matter, are crucially important to their national economies and must be able to compete in a global market. But the risk is that they dominate the rest of the urban system so the national economy becomes spatially and structurally unbalanced. Sometimes second-tier cities do benefit from national policy. But often this happens in implicit rather than explicit ways. Most states do not have a policy for second-tier cities, which means their collective interests are overlooked. The policy choice is not between favouring growing areas as opposed to the regeneration of declining areas. It is between putting the national eggs into a smaller or larger number of baskets. Our study suggests that national governments which concentrate attention and resources on their capital cities risk increasing uneven development with whole regions and cities missing out on chances to enter the new economy. Second-tier cities, although less able to act on the global stage, can still generate important dynamism for regions outside the capital and contribute to overall national growth. In many cases they punch beyond their weight. They cater for variations within nation states and contribute to territorial cohesion. They contain higher order services and offer companies better access to them than if they were all concentrated in the capital city. They can achieve many of the agglomeration effects of capitals, provided they have the right infrastructure, facilities, capacity and powers. They can lift the performance of their regions, reduce inter-regional inequalities and promote social cohesion.

4.4. *Win–Win Not Zero Sum*

So the message is clear. Strong capitals matter to nation states' global positioning and competitiveness. However, strong second-tier cities also matter. Both capital and second-tier cities must be supported in future. It is a win–win, not a zero sum relationship. A key policy issue is how to encourage second-tier cities to absorb some of their capital city's growth as capitals reach the limits of their capacity to accommodate that growth and the costs begin to outweigh the benefits. Government should help second-tier cities so they can emerge from the current recession with more "investment ready" places to maximize future national economic performance.

4.5. *Will the Market Deliver?*

We argue that second-tier cities could contribute more if they were given greater support and investment. Some argue there is no need for government intervention to address regional and urban imbalances. They believe the market itself will self-regulate and lead to increased investment in second-tier cities as the costs and price of growth in the capital become more obvious and the opportunities in second-tier cities become equally obvious. But our analysis, in keeping with much regional economic analyses, does not support that view. The logic of over-investment in the capitals and under-investment in second-tier cities is simply too strong in too many countries. As the German experience demonstrates, it requires public intervention and good governance.

4.6. *Good Governance—Shared Responsibilities and Clarity of Roles*

The key governance issue is not simply the division of powers and resources. It is also the extent to which responsibilities are shared and roles are transparent or confused. For example, urban policies tend to be vertically integrated in German cities because key functions are shared, or because the Federal Government funds urban and regional partnership experiments or because they are the subject of extensive negotiations between federal, state and city governments. Cities' financial capacity, in particular the extent to which they rely upon national grant, transfers and financial equalization or can raise their own revenue, also affects national policy impacts. In some cases, centralization of power is exacerbated by the lack of strong, democratically elected regional government and fragmented metropolitan governance. In other cases, cities in decentralized states were in virtually the same position as those in centralized unitary states because decentralization of responsibilities has not been matched by the decentralization of financial resources.

4.7. *Good Governance—Local Discretion, Shared Values, Flexibility and Trust*

National policies work best where there is collective understanding at different government levels of how different interventions affect cities and the right levers to pull to maximize performance. National policies are most effective where there is scope to shape them to local circumstances. This requires multi-level governance as well as human and fiscal capacity and autonomy at the city level. Also the consistency, transparency and reliability of national policy are critically important because urban economic development is a long-term business. Finally, the most robust policy systems are underpinned by a set of shared principles and values. These include: focussing upon business and community needs; understanding and responding to future urban challenges; reconciling strategic and local perspectives; trust, reciprocity and mutual respect.

4.8. *Territorial Economic Governance at Scale*

Few countries or cities have successfully addressed the key territorial challenge of developing economic governance at scale so that all the key actors and institutions across a functional economic area can maximize their assets to achieve integrated, place-based economic strategies. Too many cities are still attempting to use nineteenth century local boundaries and twentieth century forms of government to shape and develop a twenty-first century global economy. Successful city-regions need governance to be upscaled to the functional economic level. At the moment too often, they are too many, too small and not fit for purpose. European, national and regional governments should incentivize and encourage voluntary collaboration but also strengthen formal territorial governance at the city region level.

4.9. *Greater Transparency about Territorial Investment Strategies. Greater Focus on Second-Tier Cities*

Urban policy across Europe is very uneven. There has been a shift in the orientation of explicit urban policies and greater emphasis on boosting urban competitiveness. But the national and regional funds allocated for them are dwarfed by mainstream spending programmes. Few states consider the effects of mainstream programmes and spending on the performance of second-tier cities, since most governments are organized on functional rather than territorial lines. Also, very few states have introduced conscious policies to promote their leading second-tier cities. Governments should be more transparent about their criteria for territorial investment and their impacts upon different city-regions. Governments should monitor and publicize the territorial impacts of their expenditure programmes. In particular, Governments should ensure that all mainstream programmes, as well as special urban programmes are focussed on second-tier cities and not concentrated upon the capitals. National government policies, for example, for innovation, research and development, education and skills, transport and connectivity, and infrastructure investment have a major impact upon the relative performance of capital and second-tier cities. It is crucial they are used strategically to avoid over concentration upon, and overheating of, the capital as well as to avoid the limiting of scarce resources to second-tier cities. These principles will become more significant in a period of austerity.

4.10. *So When Should National Governments Invest in Second-Tier Cities?*

The number of high-performing second-tier cities a country can sustain will vary according to both the country's size and level of economic development. For example, in smaller countries there will be less scope for a large number of cities to complement the capital. Equally in the developing economies of the east, at present the capital city is the most significant driver of the national economy. In both cases, capital cities might remain the initial focus for investment because they are most likely to have the capacity and critical mass to succeed. Nevertheless, countries must have strategies for developing second-tier cities, to spread economic benefits and help them become the economic motors of their wider regions. Governments should encourage as many successful second-tier cities as the population and pattern of economic growth and development permit.

So for policy-makers at all government levels the message is clear. Strong capitals matter to nation states' global positioning and competitiveness. However, strong second-tier cities also matter. Both capital and second-tier cities must be supported in future. It is a win–win, not a zero sum relationship. A key policy issue is how to encourage second-tier cities to absorb some of their capital cities' growth as capitals reach the limits of their capacity to accommodate that growth, and the costs begin to outweigh the benefits. Governments at all levels should help second-tier cities so they can emerge from the current recession with more "investment ready" places to maximize future national economic performance. The individual circumstances of countries, regions and city-regions will vary and so will policy responses. However, some general principles to guide future territorial investment have become clear. Specifically, governments should invest more in second-tier cities when: (i) the gap with capitals is large and growing; (ii) the business infrastructure of second-tier cities is weak because of national underinvestment; and (iii) there is clear evidence about the negative externalities of capital city growth.

4.11. *Policy Messages for Europe*

A final policy message is for the European Commission. City-regions are crucial to the delivery of its strategic goals identified in EU2020 (European Commission, 2010). It must take city-regions—and their leadership—more seriously in future. Commission policy for cities has varied in recent years and the significance of the economic place-making agenda has fluctuated. The issues have slipped down the Commission's agenda in recent years and should be reasserted. The Commission needs to exercise leadership and provide clarity and resources in this field. It should do more to ensure that the economic potential of second-tier cities is clearly recognized in its strategies. The territorial impact of all Commission policies, not just those of DG Regio should be made more explicit. The sectoral policies of the Commission should be better integrated. But the key challenge is to ensure that not only the explicit targeted resources but all mainstream Commission funding impacts on second-tier cities in a more coherent way than it currently does. In a period of austerity, it is crucial that the Commission commits to the importance of those cities. First it should not retreat to a policy of concentrating only on small socially deprived areas but focus more widely upon economic place making. Second it must not focus only on a limited number of already successful places but should make the wider longer term investments that will bring longer term economic prosperity to more places, more countries and hence to Europe.

References

Amin, A. & Thrift, N. (1995) Globalization institutional thickness and the local economy, in: P. Healey, S. Cameron, S. Davoudi, S. Graham & A. Madanipour (Eds) *Managing Cities: The New Urban Context*, pp. 91–108 (Chichester: Wiley).

Ascani, A., Crescenzi, R. & Iammarino, S. (2012) *Regional Economic Development: A Review*, SEARCH Working Paper 1/03 (London: Department of Geography and Environment London School of Economics). Available at http://www.ub.edu/searchproject/wp-content/uploads/2012/02/WP-1.3.pdf (accessed 26 July 2013).

Barca, F. (2009) *An Agenda for a Reformed Cohesion Policy: A Place Based Approach to Meeting European Union Challenges and Expectations* (Brussels: DG Regio). Available at http://ec.europa. eu/regional_policy/archive/policy/future/pdf/report_barca_v0306.pdf (accessed 26 July 2013).

Barca, F. & McCann, P. (2010) The place-based approach: A response to Mr Gill. Available at http://www.voxeu. org/article/regional-development-policies-place-based-or-people-centred (accessed 26 July 2013).

Barca, F., McCann, P. & Rodríguez-Pose, A. (2012) The case for regional development intervention: Place-based versus place-neutral approaches, *Journal of Regional Science*, 52(1), pp. 134–152.

Brenner, N. & Keil, R. (Eds) (2006) *The Global Cities Reader* (Abingdon: Routledge).

Crescenzi, R. & Rodríguez-Pose, A. (2011) Reconciling top-down and bottom-up development policies, *Environment and Planning A*, 43(4), pp. 773–780.

Duranton, G. & Puga, D. (2004) Micro-foundations of urban agglomeration economies, in: V. Henderson & J.-F. Thisse (Eds) *Handbook of Regional and Urban Economics*, Vol. 4, pp. 2063–2117 (Amsterdam: North-Holland).

European Commission (2010b) *EU2020* (Brussels: European Commission).

Fujita, M., Krugman, P. & Venables, A. (1999) *The Spatial Economy* (Cambridge, MA: MIT Press).

Garcilazo, J. E., Martins, J. O. & Tompson, W. (2010) Why Policies may need to be place-based in order to be people-centred. Available at http://www.voxeu.org/index.php?q=node/5827 (accessed 26 July 2013).

Grabher, G. (Ed) (1993) *The Embedded Firm: On the Socio-Economics of Industrial Networks* (London: Routledge).

Henderson, J. V. (2010) Cities and development, *Journal of Regional Science*, 50(1), pp. 515–540.

Hildreth, P. & Bailey, D. (2013) The economics behind localism in England, *Cambridge Journal of Regions Economy and Society*, 6(2), pp. 2233–2249.

Kitson, M., Martin, R. & Tyler, P. (2004) Regional competiveness: An elusive yet key concept, *Regional Studies*, 38(9), pp. 9991–999.

Krugman, P. (1990) *Rethinking International Trade* (Cambridge, MA: MIT Press).

Krugman, P. (1991) *Geography and Trade* (Leuven: Leuven University Press).

Krugman, P. (1993) On the relationship between trade theory and location theory, *Review of International Economics*, 1(2), pp. 110–122.

Markusen, A., Lee, Yong-Sook & DiGiovanni, S. (1999) *Second Tier Cities: Rapid Growth Beyond the Metropolis* (Minneapolis: University of Minnesota Press).

Maskell, P. (2002) Social competitiveness innovation and competitiveness, in: S. Barron, J. Field & T. Schuller (Eds) *Social Capital: Critical Perspectives*, 111–123 (Oxford: Oxford University Press).

McCann, P. & Ortega-Argilés, R. (2011) *Smart Specialisation Regional Growth and Applications to EU Cohesion Policy*, Economic Geography Working Paper 2011, Faculty of Social Sciences, University of Groningen. Available at http://www.rug.nl/staff/p.mccan/McCannSmartSpecialisationAndEUCohesionPolicy (accessed 26 July 2013).

McCann, P. & Rodríguez-Pose, A. (2011) Why and when development policy should be place-based, in: OECD (Ed) *OECD Regional Outlook 2011: Building Resilient Regions for Stronger Economies*, 203–213 (Paris: OECD).

OECD (2006) *Competitive Cities in the Global Economy March* (Paris: OECD).

OECD (2012a) *Promoting Growth in All Regions: Lessons from Across the OECD*, March (Paris: OECD Policy Brief).

OECD (2012b) *Promoting Growth in All Types of Regions* (Paris: OECD).

Parkinson, M., Meegan, R., Karecha, J., Evans, R., Jones, G., Tosics, I., Gertheis, A., Tönko, A., Hegedüs, J., Illés, I., Sotarauta, M., Ruokolainen, O., Lefèvre, C. & Hall, P. (2012) *Second Tier Cities and Territorial Development in Europe: Performance Policies and Prospects Executive Summary and Final Report* (Luxembourg: ESPON). Available at www.espon.eu (accessed 1 April 2014).

Parkinson, M. & Meegan, R. (2013) Economic place making: policy messages for European cities, *Policy Studies*, 34(3), pp. 377–400.

Parkinson, M., Meegan, R. & Karecha, J. (2014) *UK city-regions in growth and recession: How are they performing at home and abroad?* ESRC Secondary Data Analysis Initiative Project Working Paper, Liverpool, European Institute for Urban Affairs, Liverpool John Moores University.

Piore, M. & Sabel, C. (1984) *The Second Industrial Divide: Possibilities for Prosperity* (New York: Basic Books).

Sassen, S. (2001) *The Global City: New York London Tokyo*, 2nd ed. (Princeton, NJ: Princeton University Press).

Sassen, S. (2012) *Cities in a World Economy*, 4th ed. (Thousand Oaks, CA: Pine Forge Press).

Scott, A. J. (1988) *New Industrial Spaces* (London: Pion).

Storper, M. & Scott, A. J. (1988) The geographical foundations and social regulation of flexible production complexes, in: J. Wolch & M. Dear (Eds) *The Power of Geography*, 19–40 (Boston, MA: Allen & Unwin).

World Bank. (2009) *World Development Report 2009: Reshaping Economic Geography* (Washington, DC: World Bank).

The Rise of Second-Rank Cities: What Role for Agglomeration Economies?

ROBERTO CAMAGNI, ROBERTA CAPELLO & ANDREA CARAGLIU

ABC Department, Politecnico di Milano, Milan, Italy

ABSTRACT   *In the last 15 years, empirical evidence has emerged about the fact that European first-rank cities have not always led national economic performance, and when they did, the difference between first- and second-rank cities in explaining national growth has not been significant. A recent work [Dijkstra, L., Garcilazo, E. & McCann, P. (2013) The economic performance of European cities and city regions: Myths and realities, European Planning Studies, 21(3), pp. 334–354] claims that second-rank cities have in fact outperformed first-rank cities, becoming the main driving forces in national economic performance. In the debate that emphasizes the role of second-rank cities in national growth, a simplified view of the role of agglomeration economies is provided; they are taken for granted in small- and medium-sized cities and only in large cities will the problem of a downturn in urban returns to scale emerge. In this paper, a more complex view is assumed, claiming that the oversimplified interpretation that urban economic performance simply depends on the exploitation of agglomeration economies and that these agglomeration economies merely depend on urban size alone should be abandoned. Some already existing theoretical frameworks in urban economics can help in recalling the role of possible bifurcations in the development path of cities, linked to the capability to attract or develop new and higher-order functions, increase internal efficiency and reach scale economies through cooperation networks with other cities (the city-network theory). All these elements work as conditions for fully exploiting agglomeration economies and ways to overcome urban decreasing returns.*

1. Introduction

The interest in the role of cities in explaining national economic performance is not at all new and has always been at the basis of urban economics theory and economic geography studies. In the 1990s, thanks to seminal contributions (Glaeser *et al.*, 1992; Krugman, 1991), a resurgence of interest in cities and in their role in national economic performance was registered, followed by more recent studies, especially on the North American reality (Glaeser, 2008; Henderson, 1974, 1985, 1996; Rosenthal & Strange, 2004; Sassen, 2002; Scott, 2001).

This revival of interest in the role of cities in explaining national economic performance is not merely driven by an academic fashion (Henderson, 2010), but finds concrete evidence of real changes in the role of large cities in driving national economies (Nijkamp & Kourtit, 2011, 2012). For a long time, in the last 20 years of the last century, large European cities (but not only European) benefitted from two main favourable exogenous factors that enhanced their generalized performance: the emergence and consolidation of the ICT paradigm, that was quickly utilized to re-launch urban activities after the previous de-industrialization period, and the birth of the European Single Market project by president Jacques Delors in 1985, which generated an enormous inflow of FDI into large gateway cities in all European countries (Camagni, 2001). On the other hand, the economic crisis of more recent years mostly hit large and capital cities, natural loci of core economic and financial activities. This implies that second-rank cities have proved to be the most resilient areas to economic downturn in advanced economies.

The evidence of a scarcity of public resources sharpened the debate on the contribution that each territory can provide to national competitiveness, encouraging a comparison between the efficiency displayed by capital cities vs. second-rank cities in exploiting public investment for growth. It is in fact well known that large cities are expensive machines, requiring large social overhead capital investments and exhibiting expensive real estate markets and high-rise, capital-intensive buildings. Second-rank cities, once endowed with some necessary preconditions for a modern development—namely international links, high-education and cultural facilities—may well exhibit higher public resources efficiency and better quality of life conditions than first-rank cities, being in a condition to find appropriate specialization niches inside the international division of labour.

The vast academic and policy debate over the last years has generated some important empirical studies comparing the role of second- vs. first-rank cities in explaining national economic competitiveness (Glaeser, 2011). While much US literature celebrated the role of large metro areas in fostering economic growth, there is common evidence that in Europe, over the last two decades, second-rank cities have often outperformed first-rank cities, or, even when first-rank cities grew faster, the difference with respect to second-rank cities has been negligible. In fact, in several countries, second-rank cities have been identified as the main driving forces in national economic performance (Dijkstra *et al.*, 2013; Parkinson *et al.*, 2014).

In this literature, there is a simplified view of the role of agglomeration economies; they are taken for granted in small and medium-size cities and only in large cities will the problem of a downturn in urban returns to scale eventually emerge.

Here, a more complex view is assumed. In fact, this paper adds to the existing literature the idea that urban economic performance does not only depend on the exploitation of agglomeration economies and that, in turn, agglomeration economies do not only depend on urban size alone. Some existing theoretical frameworks in urban economics can help recall the role of possible bifurcations in the development path of cities, linked to the capability to attract or develop new and higher-order functions (the SOUDY model, Camagni *et al.*, 1986), increase internal efficiency and reach scale economies though cooperation networks with other cities (the city-network theory, Camagni, 1993; Conti & Dematteis, 1995). All these elements work as conditions for fully exploiting agglomeration economies and ways to overcome urban decreasing returns.[1]

This paper aims at inspecting the conditions leading second-rank cities in the EU in some periods to outperform with respect to larger metro areas. Urban rank is defined according to the cities' physical (population) size; in particular, first-rank cities are defined as those cities (LUZ areas) with a population larger than 1 million inhabitants. Second rank cities are instead identified as those LUZ areas with a number of inhabitants in a range from 1 million to 200,000 inhabitants. In both cases, data used to define the classes refer to 2011.[2]

The interpretation assumed in this work is that cities may experience a halt in their growth path and even a decline irrespective of their size class, in the absence of these conditional factors. These factors are not really quantitative in nature, but rather qualitative and some quantum jumps in their endowment are needed at specific intervals if agglomeration economies have to fully generate their beneficial effects. The quality of activities hosted, the quality of production factors, the density of external linkages and cooperation networks, the quality of urban infrastructure—in internal and external mobility, in education, in public services—are all enabling factors allowing a long-term 'structural dynamics' process (in the language of dynamic modelling) through what could easily be called a process of urban innovation in each urban category.

The paper thus departs from most existing literature, by decomposing the black box of agglomeration economies, and their impacts on long-run urban performance, into their main determinants, as well as by better qualifying the commonplace finding that agglomeration economies/diseconomies would be the only determinants of urban dynamics.

The paper is structured as follows. Section 2 briefly presents the recent debate on the rise of the second-rank cities, and highlights the existing conceptual toolbox that allows overcoming the traditional view of the role of agglomeration economies in urban growth. Section 3 presents a descriptive analysis on the economic performance of the different size classes. Section 4 presents the conceptual model and the rich database on which the model is estimated. Section 5 contains the results of the econometric estimates, while Section 6 concludes.

2. The Theoretical Explanations for the Rise of Second-Rank Cities: Testable Assumptions

Against the much celebrated triumph of (large) cities (Glaeser, 2011), recent empirical evidence has been presented on the fact that in the EU the last two decades witnessed a relatively comparable performance across cities of first and second rank. Within this recent debate, agglomeration economies are called upon to explain the relatively better performance of second-rank cities, while diseconomies of scale are identified as the cause of the limited success of large ones (Dijkstra *et al.*, 2013). The explanations provided for such a phenomenon are not convincing, and risk an ex-post circular reasoning: a structural break—namely the limit between economies and diseconomies—is used to explain cyclical economic phenomena, with no interpretation of why it takes place exactly at a specific moment and in a particular place.

In the above reasoning, two problems are mixed together: a structural one and a cyclical one. The former deals with the existence of a theoretical explanation concerning the way and conditions for the different city-size classes to exploit increasing returns to urban scale (the how problem); the latter problem concerns the question whether different classes of urban size are better suited for expansion or crisis periods (the when). The how issue is logically propaedeutical and more interesting than the when issue, as descriptive evidence shows that there are some large cities still able to play a role in their national economies, and some second-rank cities still lagging behind, demonstrating that, for the same urban size, agglomeration economies seem to play a positive role for some cities and not for others.

Urban economists are called to give some more appropriate explanations to these phenomena. In our opinion, some relevant theoretical interpretations concerning increasing returns to urban scale—going beyond the use of a single, smooth function for the entire spectrum of city sizes, interpreting the capacity of cities to exploit agglomeration economies—are already available in urban economic theory, and deserve empirical demonstration.

The first explanation for the capacity of a city to overcome diseconomies of scale is contained in the so-called SOUDY (Supply Oriented Urban DYnamics) model (Camagni *et al.*, 1986). The model assumes that an "efficient" city-size interval exists separately for each hierarchical rank, associated with rank-specific economic functions. In other words, for each economic function characterised by a specific demand threshold and a minimum production size, a minimum and a maximum city size exists beyond which urban location diseconomies overcome production benefits typical of that function.

As Figure 1 shows, under these conditions, for each economic function and each associated urban rank, it is possible to define a minimum and a maximum city size in which the city operates under efficiency conditions (i.e. with net positive gains) ($d_1 - d_2$ for the function—and centre—of rank 1; $d_3 - d_4$ for the function—and centre—of rank 2; ...). The higher the production benefits (profits) of the single functions (increasing with rank), the higher the efficient urban size interval associated to such function.

As each centre grows, approaching the maximum size compatible with its rank ("constrained dynamics"), it enters an instability area (e.g. in $d_3 - d_2$ in Figure 1) where it becomes a potentially suitable location for higher-order functions, thanks to the achievement of a critical demand size for them. In dynamic terms, each city's long-term growth possibilities depend on its ability to move to higher urban ranks, developing or attracting

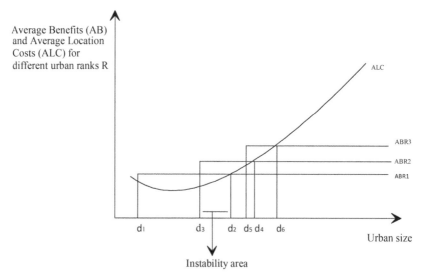

Figure 1. Efficient city size for different urban functions.
Source: Camagni *et al.* (1986).

new and higher-order functions ("structural dynamics"). This "jump" is not mechanically attained: it represents a true urban innovation and is treated as a stochastic process in the dynamic model.

The interest of this model resides in the fact that it overcomes some of the limits of the "optimal" city-size theory, by suggesting:

- the need to replace "optimal size" by a "range" within which the city size is "efficient",[3] i.e. where average production benefits exceed average location costs;
- the need to allow different "efficient" urban intervals according to the functions actually performed by the cities;
- the possibility of decoupling urban ranks from urban size. Differently from Christaller's approach, two cities of the same size (for example, size d_2 in Figure 1) can belong to two ranks (1 and 2 in the example), depending on their capacity to attract/develop higher functions.[4]

The second theory helpful in explaining the determinants of agglomeration economies that go beyond urban size is the city-network theory. Born in the field of industrial economics (Chesnais, 1988; Williamson, 1985), the concept of network behaviour—namely a cooperative organizational form, intermediate between internal and external growth of the firm, between "make" or "buy"—was transferred into urban economics providing a successful theoretical framework to overcome the limiting interpretative power of the traditional central-place model.[5] In fact, real city-systems in advanced countries have departed from the abstract Christaller pattern of a nested hierarchy of centres and markets, showing (Camagni, 1993):

- processes of city specialization and presence of higher-order functions in centres of lower order;
- horizontal linkages between similar cities, not allowed in the traditional model: e.g. the financial and headquarter networks among top world cities, or the art cities linked through tourist "itineraries" (so called "synergy networks"); the systems of specialised cities exploiting the advantages of the division of labour, like the Dutch Randstad or the network of medium-sized cities in the Veneto region ("complementarity networks") (Camagni & Capello, 2004).

In the new logic, other elements come to the fore—economies of vertical and horizontal integration, and network externalities similar to those emerging from "club goods". These elements provide the possibility for cities to reach increasing returns and scale economies through network integration—in the economic, logistic and organizational fields—with other cities.

In this sense, the concept of city networks recalls that of "borrowed size", which was launched by Alonso (1973) to explain a disconnection between size and function of smaller cities that were part of a megalopolitan urban complex:

> the concept of a system of cities has many facets, but one of particular interest ... is the concept of borrowed size, whereby a small city or metropolitan area exhibits some of the characteristics of a larger one if it is near other population concentrations. (Alonso, 1973, p. 200)

According to this literature, "borrowed size" can easily explain why second-rank cities grow economically without a physical expansion: their location in polycentric urban systems substitutes for their individual physical size (Meijers & Burger, 2010), since borrowed size allows single cities to upgrade their economic functions without necessarily increasing their individual size. Externalities accrue to functionally connected urban areas, that thus reach economies of scale without incurring the costs associated with (excessive) urban size (Phelps *et al.*, 2001; Veneri & Burgalassi, 2012; and, for a conceptual review, Parr, 2002). Conversely, a polycentric urban structure, whereby cities of average size can overcome the limits associated with sheer size, may be associated with limitations to the achievement of economies of scale whenever the threshold of given services cannot be dispersed in such a spatial arrangement (see for instance Burger *et al.*, 2013, 2014).

An important difference, however, exists between the concept of borrowed size and that of city networks. The concept of city networks adds to that of "borrowed size" the idea that size can be borrowed not only thanks to physical proximity to larger centres, but also thanks to relationships and flows of a mainly horizontal and non-hierarchical nature among complementary or similar centres, located far from each other, with the aim of achieving network externalities (Camagni, 1993; Camagni & Capello, 2004; Capello, 2000).

These conceptual ideas help in explaining why cities of intermediate size are being increasingly looked upon as the places that could well host the growth of the years to come: limited city size, in fact, facilitates environmental equilibrium, efficiency of the mobility system and the possibility for citizens to retain a sense of identity, provided that a superior economic efficiency is reached through external cooperation with other cities, located in the same regions or distant but well connected. Urban productivity was empirically found to be much more closely related to urban connectivity—a concept similar to urban network relations—than to urban scale (McCann & Acs, 2011), thus supporting the global city argument but also the medium-sized cities' potentials.

The joint application of the SOUDY model and the city-network paradigm has relevant implications for urban efficiency and growth: size is not the only determinants of factor productivity and urban performance. The presence of high-order functions and integration inside city networks are also extremely important elements in the explanation of the competitive advantage of cities, allowing the boosting of productivity even in presence of limited urban size.

We expect functional specialization and city networks to play a prominent role in explaining the achievement and exploitation of agglomeration economies (or in avoiding agglomeration diseconomies) in second-rank cities.[6] Moreover, within second-rank cities, we expect best performing cities to be able to exploit increasing returns much more than the other cities.

Our research hypotheses are therefore the following:

(a) small- and medium-sized cities can also experience a halt or a decline in their growth pattern, since each city's long-term growth possibilities depend on its ability to move to higher urban ranks, developing or attracting new and higher-order functions;
(b) the best performing small- and medium-sized cities are able to postpone and overcome the appearance of decreasing returns, thus being able to fully exploit increasing returns to urban scale;
(c) the capability of best performing small- and medium-sized cities to exploit increasing returns depends on the development of higher valued functions and external cooperation networks;

(d) For second-rank cities, these growth-enhancing factors can be borrowed from closely located large urban areas.

3. Urban Economic Performance from 1996 to 2009: Second- vs. First-Rank Cities

This section presents a comprehensive picture of the medium-run trends for a sample of 136 EU metropolitan areas, dividing the trends between first-rank and second-rank cities. The two "uniform categories" of first- and second-rank cities are defined in terms of physical size, as usually done in the economic geography literature. As anticipated in Section 1, first-rank cities are defined as those cities (LUZ areas) with a population larger than 1 million inhabitants. Second rank cities are instead identified as those LUZ areas with a number of inhabitants in a range from 1 million to 200,000 inhabitants. In both cases, data used to define the classes refer to 2011.

Figure 2 presents the 1996–2009 time series for yearly gross domestic product (GDP) growth rates divided by urban rank. A first result is that a clear cyclical process character-ized European economies. Figure 2 displays annual GDP growth rates on the main graph, while showing aggregate GDP growth rates in the bottom (first through third period) and top-right (fourth and last period) corner of the figure.

The main graph suggests a few relevant stylized facts for the EU economies in the sur-veyed period:

• the EU economy, following a worldwide trend, behaves cyclically. Two fast growth stages can be identified, running from 1995 to 2000 and from 2003 to 2007; between such

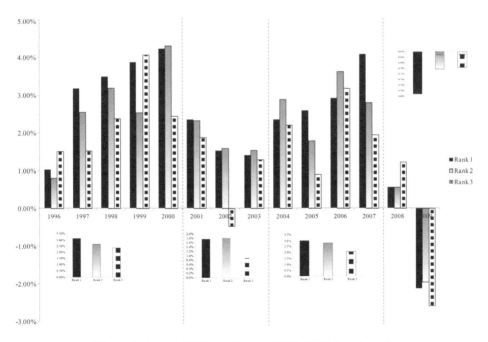

Figure 2. Annual GDP growth rates, 1996–2009, by city rank.
Note: In the boxes: period averages.
Source: authors' elaboration.

periods, three years of growth slowdown (2000–2003) can be identified. Finally, after 2007 EU data clearly show the appearance of the (currently ongoing) financial crisis, first marking a relevant growth slowdown in 2008, then showing the first signs of GDP contraction in 2009. In Figure 2, these periods are marked with vertical dashed lines;

- most interestingly, while in the fastest expansion periods rank 1 cities seem to drive economic growth (black-coloured bars in Figure 2), in slowdown times rank 2 cities (shaded colour in Figure 2) invert this trend and are at the forefront of the process of economic development. In Figure 2 this statement can be easily seen from the ancillary graphs at the bottom and top-right corner of the main chart, showing period average growth rates. Rank 2 cities grow in fact faster than rank 1 cities in the second period, and present a less pronounced decline in GDP in the fourth period;

- while this EU-wide trend is relatively visible with plain GDP growth data, the prevalence of rank 2 cities as engines of economic growth in recent periods emerges more clearly from the inspection of Figure 3(a)–(c). Figure 3, in fact, represents per capita GDP growth rates for the same four periods, showing that indeed in 8 of the last 10 years rank 2 cities outperformed rank 1 cities, thus contributing more to the overall country (and EU-wide) growth of productivity. This statement is again valid for both the second as well as for the last period above identified;

- while Figure 3(a) shows such trend for the whole EU27, Figures 3(b) and 3(c) break down this result by macro areas. Figure 3(b) shows data for productivity growth in the EU15, and Figure 3(c) for New Member States (henceforth, NMS). Because of the large portion of EU27 GDP produced by Western Countries, Figure 3(b) behaves quite similarly to the main EU27-wide trend, with rank 2 cities growing faster than rank 1 cities in the second and fourth period. Figure 3(c) shows instead that in most observed years rank 1 cities overperform with respect to rank 2 ones in NMS, with no clear trend of a rank 2 cities' reprise. Clearly, this may depend either on a strong prevalence of rank 1 cities in the process of wealth creation in NMS, or else on the relative lack of a well-structured, well-interconnected network of rank 2 cities in these countries.[7]

Graphical evidence does not find, however, unequivocal statistical support in classic t-tests for mean differences across groups, and this holds for all the four periods identified. This result strengthens the case for our empirical analyses, as it clearly shows that first-rank cities are not really leading national performance as superficially expected on the basis of their size and role in national economies, and points at the need to analyse the reasons why some second-rank cities have outperformed first-rank ones in the last 20 years. Thus, this work focuses on the two periods (2001–2004 and 2008–2009) when rank 2 cities grew faster than first-rank cities. In these periods, a dummy variable is calculated for second-rank cities, taking on value 1 if the metro area overperformed with respect to the country average GDP growth rate. The four periods identified in the data are shown in Table 1.

4. The Model and the Data Description

4.1. The Estimated Model

The research questions mentioned in Section 2 are addressed by estimating an aggregate urban production function in the traditional neoclassical form of average location benefits

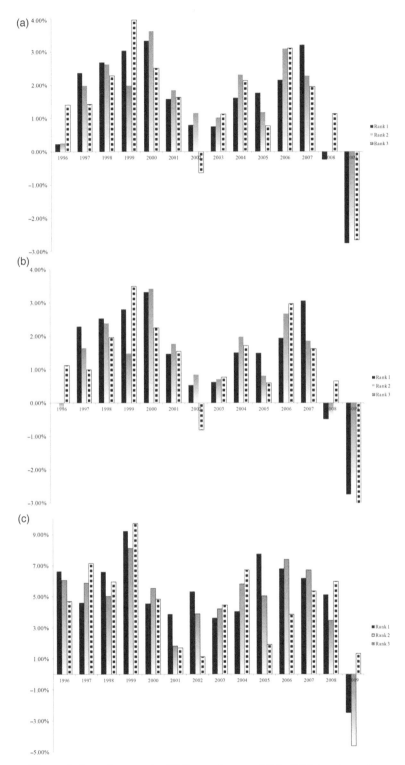

Figure 3. Annual per capita GDP growth rates, 1996–2009, by city rank.
(a) EU27. (b) EU15. (c) NMS.
Source: authors' elaboration.

Table 1. Periods observed in the data set used for the empirical analyses

Period	Label	Period observed in the dependent variable	Period observed in the explanatory variables
1995–2001	Growth	–	–
2001–2004	Slowdown	2004	Av. 1998–2002
2004–2008	Reprise	–	–
2008–2009	Crisis	2011	Av. 2002–2006

(henceforth, ALBs). ALB depend on the size of the city, measured through the absolute number of inhabitants, as most of the traditional literature developed (Catin, 1991; Marelli, 1981; Rousseau, 1995; Rousseau & Prud'homme, 1992; Segal, 1976; Sveikauskas *et al.*, 1988). In order to discern between increasing or decreasing returns to scale, a quadratic form is imposed on the ALB by introducing the square population term in the model. Moreover, networks and functions are expected to act on average benefits when they have achieved a certain critical mass of population, and this is captured by the interaction term between networks and functions, on the one hand, and population, on the other hand.

The basic model estimated therefore reads as follows:

$$\text{ALB} = \text{const} + \beta_1 \text{Pop} + \beta_2 \text{Pop}^2 + \beta_3 \text{Pop}^*\text{Functions} + \beta_4 \text{Pop}^*\text{Networks} + \varepsilon, \quad (1)$$

where "ALB" stands for average location benefit and "Pop" stands for population.

A specific indicator of net urban advantage has been used in this work, i.e. urban rent, measured through house prices per square metre.[8] Already used with the same purpose in other studies (Camagni & Pompili, 1991; Capello, 2002), this indicator is based on a crucial underlying hypothesis, i.e. that the differences in house prices between large and small cities measure their relative attractiveness (and thus their net localisation advantage), since they are the result of an evaluation made by the market of the "value" of these locations. For the same reason, the dynamics of urban house prices captures the changes in attractiveness of each location, and thus the dynamics of urban net advantage.[9]

According to this logic, urban rent is a useful indicator of advantages and costs of different urban sizes, and of the existence of an optimal city size. This latter is the size which allows the achievement of the maximum net agglomeration advantages, estimated by the market as an "equilibrium rent level". For urban areas larger than the optimal size, scale diseconomies would emerge, causing a relocation of residential and production activities, a decrease in city size and a consequent decrease in urban rent. In the same way, a city which is smaller than the optimal size would be characterised by lower rent levels, which would attract production and residential activities, causing an increase in city size and consequently an increase in urban rent.

The outperforming small- and medium-sized cities with respect to the national average have been identified, and a dummy created for them. In order to analyse specificities of the outperforming cities with respect to the average behaviour, the basic model has been estimated again by introducing the interaction terms between the independent variables and the outperforming city dummies, as follows:

$$ALB = const + \beta_1 Pop + \beta_2 Pop^2 + \beta_3 Pop^* Functions + \beta_4 Pop^* networks + \beta_5 Pop^* D+$$
$$+ \beta_6 Pop^{2*} D + \beta_7 Pop^* Functions^* D + \beta_8 Pop^* Networks^* D + \beta_9 D + \varepsilon.$$

$$(2)$$

Equation (2) contains the interaction terms with D, which represents the relatively best performing cities with respect to the national average. An alternative version of model (2) entails the use of a measure of borrowed size, that aims at capturing the possibility that second-rank cities may have in fact borrowed the use of high-level urban functions and connectivity from large metro areas, irrespective of their own (relatively) small size. This implies that borrowed size partially substitutes functions and networks; if this is the case, an alternative model can be estimated, namely:[10]

$$ALB = const + \beta_1 Pop + \beta_2 Pop^2 + \beta_3 D + \beta_4 D^* Pop + \beta_5 D^* Pop^2$$
$$+ \beta_6 Borrowedsize + \beta_7 D^* Borrowedsize + \varepsilon.$$

$$(3)$$

Before entering the estimation procedure, we present the database on which the model is estimated (Section 4.2).

4.2. The Database

In order to empirically test the research questions stated in Section 2, we assembled a novel data set which exploits the relatively recent release of metro area data from EUROSTAT. Table 2 summarizes the main sources of the data set assembled for the present analysis.

The European statistical institute provides in fact a detailed definition of metropolitan statistical areas in Europe. In particular, data on metro regions are based on an aggregation of NUTS3 administrative regions with at least 250,000 inhabitants, in turn based on LUZ. As such, the aim of such statistical classification is to "correct the distortions created by commuting" by "including the commuter belt around a city" (Dijkstra, 2009, p. 1).[11]

For these regions, we collected data on GDP in current prices and population from 1995 to 2009. As for the latter, because the early years (1995–2000) were missing from the data set, population in the NUTS3 regions forming metropolitan areas has been aggregated following the official EUROSTAT classification. Finally, GDP data has been deflated with the EUROSTAT country-specific deflator, to obtain GDP in constant 2005 prices (net of within-country price variations, which cannot at the present state be fully eliminated).

The data on GDP and population have been used to identify those cities that over—or underperformed with respect to the country mean (see the evidence presented in Section 3).

Next, the data necessary to estimate the models presented in equation (2) have been collected. As explained in Section 4.1, urban land rent represents our measure of average urban location benefits. Land rent is here measured with the price per square metre of an average quality apartment located in the Central Business District of each city in the sample. The data, collected for the years 2004 and 2011, have been deflated with the same country deflator used for correcting GDP price distortions, in order to refer to constant 2005 prices.

Moreover, data for high urban functions and city networks are also needed to test equation (2). For urban functions, the share of high-skilled professionals over the total population is used, from the micro database of EUROSTAT's labour force survey, aggregated at the

Table 2. The data set

Data	Indicator	Source of raw data	Years available
GDP	Constant 2005 € GDP	EUROSTAT	1995–2009
Population	Average yearly metro area population	EUROSTAT	1995–2009
Land rent	Prices of an average apartment in the city's CBD	EUROSTAT/Urban Audit, author's integration[a]	2004, 2011
High urban functions	Share of high-skilled professionals over total population[b]	Labour Force Survey	1995–2007 (average 1998–2002 and 2002–2006 have been calculated)
City networks	Number of FP 5 and 6 co-participations	CORDIS	1998–2002 (FP5) 2002–2006 (FP6)

Source: Authors' elaboration.
[a]House prices data have been integrated through single national sources.
[b]High-level functions are defined as the ISCO 88 category 1, encompassing "legislators, senior officials and managers". Micro data have been aggregated at NUTS2 level, and the value of the NUTS2 region is assigned to the metro area. A full list of ISCO professions is available at http://laborsta.ilo.org/applv8/data/isco88e.html.

NUTS2 level (see Table 1 for details). In order to measure urban networks, the approach used in Camagni *et al.* (2013) has been followed and urban networks are measured by the number of Framework Programme (henceforth, FP) 5 and 6 projects in which institutions of each surveyed city participated.[12] This indicator has the advantage of capturing collaborations and exchange of knowledge, irrespective of the geographical distance between network nodes. As such, they represent a good proxy for the capability of a city to engage in long-distance networks aiming at fostering scientific cooperation, and represent a comprehensive measure of the quality and thickness of external urban networks (Basile *et al.*, 2012, p. 710).

Finally, borrowed size (viz. the accessibility of second-rank cities to functions and networks typical of large urban areas) is calculated as the population in first-rank urban areas discounted by distance from each second-rank city. Borrowed size (equation (3) above) is therefore defined as:

$$Borrowedsize_i = W\overline{pop} = \sum_{j=1}^{n} pop_j w_j, \quad \forall j \neq i, \tag{4}$$

where i and j represent respectively second- and first-rank cities, W is an nXn distance weight matrix between second- and first-rank cities (each entry in the matrix is equal to 0 if the two cities are second-rank or first-rank cities and equal to the inverse of the geographical distance in kilometres if the two cities are cities one of second- and one of first-rank), and $\overline{pop}$ represents the vector of first-rank city populations.[13]

5. An Interpretative Model of Urban Structural Dynamics of Small- and Medium-Sized Cities

5.1. *Outperforming Cities: Do Agglomeration Economies Play a Role?*

Table 3 reports the results of the estimates for the empirical model, and for equation (2). The results are presented in columns 1–4 for the second-rank cities of Western countries.

Table 3. Estimation results (random effects)

Dep. variable	Average urban benefits (Log urban rent)			
Model	(1)	(2)	(3)	(4)
Constant term	−3.38	−0.72	−46.51	−40.42
	(27.12)	(22.52)	(29.62)	(28.87)
Metro area population	1.38	1.17	7.84*	7.06*
	(3.97)	(3.30)	(4.36)	(4.23)
Metro area population^2	−0.04	−0.04	−0.28*	−0.26*
	(0.14)	(0.12)	(0.16)	(0.16)
D (outperforming second-rank cities)	0.06	–	55.78*	51.29*
	(0.06)		(33.05)	(32.60)
D × metro area population	–	–	−8.47*	−7.76*
			(4.86)	(4.80)
D × metro area population^2	–	–	0.32*	0.30*
			(0.18)	(0.18)
Functions	–	0.33***	–	–
		(0.04)		
Networks	–	0.02	–	–
Functions × metro population	–	–	–	0.02***
				(0.001)
D × functions × metro population	–	–	–	0.01*
				(0.00)
Networks × metro population	–	–	–	0.003*
				(0.00)
D × networks × metro population	–	–	–	−0.004
				(0.003)
Number of obs.	158	158	158	158
Wald chi-square	6.51*	80.85***	27.20***	324.96***
Breusch–Pagan test (OLS vs. RE)	15.82***	7.05***	16.78***	7.01***
	(0.00)	(0.00)	(0.00)	(0.00)
R-squared (within)	0.07	0.29	0.13	0.37
R-squared (between)	0.04	0.43	0.06	0.44
R-squared (overall)	0.04	0.39	0.07	0.42
Robust standard errors	Yes	Yes	Yes	Yes

Notes: All variables are in log. Dependent variable: average urban benefits (Log urban rent). Functions and networks variables are expressed in logs.

Abbreviation: D—outperforming second-rank cities.

*Significance at 1%.

**Significance at 5%.

***Significance at 10%.

In fact, the model does not interpret the behaviour of all second-rank cities in Europe, testifying that Eastern European second-rank cities have still a very peculiar dynamics that has no similarities to those of advanced countries (see Section 3 for descriptive figures).

The results in columns 1–4 are obtained with random effects estimates. Fixed effects for single cities are conceptually refused in this exercise; besides, OLS estimates are statistically rejected, with the standard Breusch–Pagan test suggesting the appropriateness of random effects. Peculiarities for single cities in exploiting agglomeration advantages that do not depend on size, functions and networks neither exist, nor are expected to play a role. Also for western second-rank cities, the fit of the model is good (overall $R^2 = 0.42$).

The first column shows that the inverted U-shaped curve of the location benefit curve does not hold for the whole sample of second-rank cities in western countries, and the result does not change when functions (significant in absolute terms) and networks are added. Column 3 instead presents interesting results. When the size of the city is multiplied with a dummy representing the best outperforming second-rank cities, results become significant; while non-performing cities register a traditional inverted U-shaped form, implying an achievement of a threshold in terms of size-up to which decreasing returns are at work, fast-growing second-rank cities register decreasing returns up to a certain critical size, at which they start enjoying increasing returns. Lastly, column 4 shows the estimates for our conceptual equation (2); with the exception of the networks in outperforming large cities, the coefficients have significant and expected signs.[14]

Results are also presented in Figures. Figure 4 reports the ALB curve with respect to size. Many interesting messages are contained in this figure, namely:

- for the general sample, the curve shows an inverted U-shape relationship (the continuous curve in Figure 4). An optimal size exists, after which economies of agglomeration turn into diseconomies, and this is true also for small- and medium-sized cities, as expected according to our first assumption;
- when only the outperforming cities are analysed (dotted line in the figures), the ALB curve displays a U-shaped form, showing that increasing returns in size characterize best performing cities, i.e. agglomeration economies are associated with outperforming cities;
- outperforming cities turn diseconomies of scale that characterize cities of their size in the sample into economies, by increasing the level of functions; as Figure 4 shows, second-

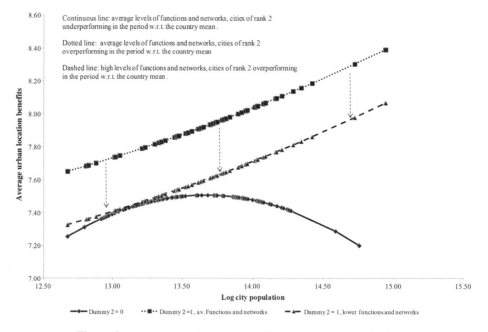

Figure 4. Average urban location benefit curve for second-rank cities.
Source: authors' elaboration.

Table 4. Optimal city size, critical size and level of functions for second-rank cities

Optimal city size (no. of inhabitants) (maximum point of the benefit curve function)	714,973
Critical size of outperforming cities turning diseconomies into economies of agglomeration (minimum point of the benefit curve function of outperforming cities)	296,559
Share of advanced functions and critical size necessary to outperforming cities to achieve the same level of urban benefit than the other cities	7.5% At a size of 545,000

Source: Authors' elaborations.

rank outperforming cities manage to achieve the bifurcation point from the average sample for lower levels of functions than first-rank cities (dashed line);
- higher intensities of high-value functions would shift the average location curve (as indicated for instance by the dotted line in Figure 4). This means that higher functions produce, ceteris paribus, higher location advantages.

Table 4 reports the optimal city size, the critical size at which, for outperforming cities, diseconomies turn into economies and the level of high functions necessary to achieve the same level of urban benefits of the other cities. Optimal second-rank city size is achieved at around 700,000 inhabitants, and requires a critical size of around 300,000 inhabitants.

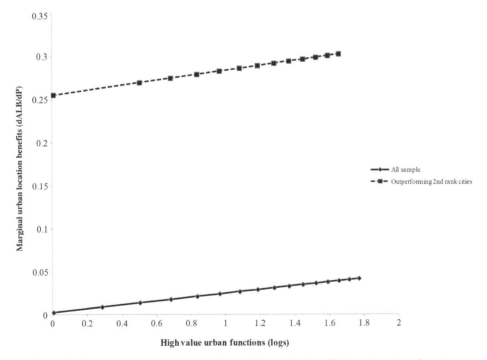

Figure 5. Marginal urban location benefits at different levels of high-value urban functions.
Source: authors' elaboration.

43

As a final consideration, second-rank cities need 7.5% of employment in high-level functions and a size of 545,000 inhabitants. Therefore, small cities require a low intensity of high-level functions and a low size to gain increasing returns, as the SOUDY model theoretically explains.

Our second hypothesis has been confirmed. Outperforming cities are the ones that are able to turn decreasing returns to scale into agglomeration economies. Moreover, increasing returns of outperforming cities are associated with higher-level functions.

5.2. Exploitation of Agglomeration Economies: Do Functions Play a Role?

The role of high-value functions in increasing the ALB curve has been presented in Figure 4. Our interest in this part of the analysis is to show if functions determine the intensity of agglomeration economies (Figure 5).[15]

Figure 5 is helpful in this respect. It shows the way in which marginal location benefits change by increasing the high-value functions. For second-rank outperforming cities the

Table 5. Estimates of borrowed size

Model	(1)	(2)	(3)
Constant term	−46.51	−59.32*	−49.95
	(29.62)	(31.48)	(31.03)
Metro area population	7.84*	9.48**	8.27*
	(4.36)	(4.54)	(4.49)
Metro area population^2	−0.28*	−0.34**	−0.30*
	(0.16)	(0.17)	(0.16)
Dummy rank 2 city outperfomed the country	55.78*	62.10*	44.09*
	(33.05)	(33.09)	(30.13)
Dummy rank 2 city outperfomed the country × metro area population	−8.47*	−9.40*	−7.24*
	(4.86)	84.87)	(4.50)
Dummy rank 2 city outperfomed the country × metro area population^2	0.32*	0.36**	0.28*
	(0.18)	(0.18)	(0.16)
Borrowed size	–	0.12	0.04
		(0.11)	(0.10)
Rank 2 city outperfomed the country × borrowed size	–		0.25*
			(0.15)
Number of obs.	158	158	158
Wald χ^2 test	4.16***	30.15***	36.19***
Breusch–Pagan test (OLS vs. RE)	16.78***	16.90***	18.01***
	(0.00)	(0.00)	(0.00)
R-squared	0.07	0.07	0.07
Within	0.13	0.15	0.18
Between	0.06	0.06	0.06
Robust standard errors	Yes	Yes	Yes
Estimation technique	RE	RE	RE

Notes: All variables are in log. Dependent variable: average urban benefits (Log urban rent). Standard errors are in parentheses.
Abbreviation: RE—random effects estimates.
*Significance at 1%.
**Significance at 5%.
***Significance at 10%.

role of functions in marginal urban benefits is high, but the intensity of the effect (the slope of the line) remains constant as the intensity of functions increases.

5.3. *Exploitation of Agglomeration Economies: Does Borrowed Size Play a Role?*

In this section, the empirical analyses presented above are complemented with a specific analysis of the role of the cities' capability of borrowing size from nearby large metro areas in fostering agglomeration economies. Table 5 shows the estimates of the baseline model, which, for convenience, is repeated in column 1, and of the model in eq. (3), which encompasses borrowed size (column 2) and the interaction term between borrowed size and the dummy for outperforming second-rank cities (column 3).

As anticipated in Section 4, Table 5 presents results where high-level urban functions and city networks are substituted by borrowed size. The results suggest that, after controlling for the other main determinants of location benefits, while borrowed size does not suffice to allow higher benefits (column 2), it acts as a catalyst for second-rank cities that outperformed with respect to the country mean (column 3). Thus, the hypothesis is that such cities can actually borrow functions and networks located in large urban areas, but without incurring the costs typically associated with locations in large urban areas.

This finding can be further evidenced graphically. This is done in Figure 6, where ALBs are plotted against the measure of borrowed size, along with confidence intervals calculated for the quintiles of borrowed size distribution.

Figure 6 suggests that indeed as second-rank cities can increasingly get access to the functions and networks hosted by large metro areas, their ALB increase even if those func-

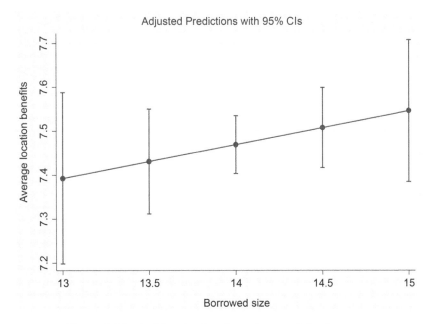

Figure 6. Marginal location benefits as borrowed size increases.
Source: authors' calculations.

tions and networks are not located within the city itself. This interesting finding paves the way for possible future research aiming at explaining the conditions for such polycentric urban structure to emerge and succeed. In fact, it could be argued that other characteristics of urban systems influence the capacity of a second-rank urban areas to profitably benefit from being close to large metro areas.

6. Conclusions

By entering the debate on the reasons for the recent relatively better performance of second-rank cities with respect to first-rank ones on the national dynamics, the present work has been able to demonstrate that some general common beliefs on the determinants of urban growth are too simplistic.

The success of cities is generally attributed to the existence of agglomeration economies and, by the same token, the urban decline is explained by the loss of increasing returns when a city achieves an excessive size. Starting from existing theoretical approaches, this paper proves that the existence of agglomeration economies is undeniable, as well as the risk of agglomeration diseconomies. However, as theoretically suggested, this paper empirically shows that cities are able to overcome diseconomies of scale either through innovating in the functions that they perform, or in the organization of activities with other cities, through city networking. Besides, such a finding is complemented by the empirical verification that cities can "borrow size" from neighbouring large metro areas, thus getting access to the functions and networks there hosted, without incurring high location disadvantages.

In fact, our empirical study has shown that outperforming cities are the ones that are characterized by economies of scale, and that these economies of scale are related to the level of functions and networks that cities have or, in other words, their capacity to borrow size from large metro areas.

These results have important urban policy implications. Concerning the recent question raised whether in a period of crisis like the present one, policy-makers should concentrate their limited resources in their large cities in order to exploit agglomeration economies, or spread their investment in a larger set of cities, the reply comes very easily after the results of this exercise. Investment should be devoted to cities in order to make any of them, irrespective of their size, be able to turn their risk of decreasing returns into agglomeration economies, by investing in renovating their functions and their way of cooperation.

Notes

1. An alternative way to explain the efficiency of second-rank cities goes through the role of size borrowing from nearby large metro areas, hosting high-level functions and being connected in transnational networks, as suggested in Alonso (1973). See Camagni *et al.* (2013).
2. More details on the empirical approach to the definition of urban rank and the relative performance of second-rank cities with respect to larger metro areas in the last 15 years are provided in Section 3.
3. Richardson (1972) suggests replacing the concept of optimal city size with an efficient interval of urban size in which urban marginal benefits are greater than marginal location costs.
4. The two cities will differ, though, in dynamic terms: the one belonging to the lower rank 1 will not grow further, having reached the maximum size of its interval, while the one having developed the higher functions (linked to rank 2) will grow, due to the presence of new and wide net urban benefits (profits).

5. Camagni (1993) theorised the concept applying it to urban systems. The same concept was already utilised in other fields, such as the behaviour of the firm and microeconomic organisational behaviour. For a review of the concept, see Capello and Rietveld (1998).

6. In another paper, the same authors show that these features are common to cities of different size, demonstrating that cities grow according to the same structural laws, but with some specificities (Camagni *et al.*, 2013).

7. For this reason, in the empirical analysis (Section 5), rank 2 cities will only be analysed in EU15 countries. The trends above discussed allow the identification of four periods within the 1995–2009 time span observed in this paper.

8. Urban rent is usually interpreted as the rent paid to the house owner. However, house prices represent the capitalized rent over time, and for this reason may be chosen as a proxy for urban rent.

9. In dynamic terms, the reasoning requires another important hypothesis. Since the analysis is developed in relative and not absolute terms, between different cities or between core and ring areas, it is assumed that for each relative dimension (large vs. small cities, ring vs. core), the supply curve of houses has the same slope. If this were not the case, a shift upwards of the demand curve, generated by a higher appreciation of location advantages, would give rise to a different increase in prices. This hypothesis does not limit too much the comparison between large and small cities, but could give a heavy bias in a comparison between core and ring areas because of the different potential for the expansion of residential supply in the two areas.

10. Interestingly enough, when a measure for borrowed size is inserted in our estimated model, multicollinearity exists among borrowed size from one side, and functions and networks on the other.

11. A full list of cities surveyed in this paper is available in the appendix.

12. Clearly, this indicator represents only a subset of all possible ways of transnational networking for the surveyed urban areas, and in particular refers to a scientific type of connectivity (thus reflecting an innovation-oriented type of knowledge being exchanged via these connections). However, this is perfectly in line with the scope of this analysis, and fits much better the aims of the empirical work here presented, rather than the use of more classical (and geographically based) networks of physical accessibility.

13. The bar notation indicates a vector.

14. The non-interacted variables "networks" and "functions" have not entered this equation; lacking degrees of freedom, we preferred to keep the specification that is conceptually the most interesting for us.

15. Networks turned out to be insignificant, and for this reason they are not treated in this part. Their role will be highlighted when a diachronic analysis is developed. See Section 5.3.

References

Alonso, W. (1973) Urban zero population growth, *Daedalus*, 102(4), pp. 191–206.

Basile, R., Capello, R. & Caragliu, A. (2012) Technological interdependence and regional growth in Europe: Proximity and synergy in knowledge spillovers, *Papers in Regional Science*, 91(4), pp. 697–722.

Burger, M. J., Meijers, E. J. & Van Oort, F. G. (2013) Regional spatial structure and retail amenities in the Netherlands, *Regional Studies*, doi: 10.1080/00343404.2013.783693

Burger, M. J., Meijers, E. J., Hoogerbrugge, M. M. & Tresserra, J. M. (2014) Borrowed size, agglomeration shadows and cultural amenities in North-West Europe, *European Planning Studies*, http://dx.doi.org/10. 1080/09654313.2014.905002.

Camagni, R. (1993) From city hierarchy to city networks: Reflection about an emerging paradigm, in: T. Lakshmanan & P. Nijkamp (Eds) *Structure and Change in the Space Economy: Festschrifts in Honour of Martin Beckmann*, pp. 66–87 (Berlin: Springer Verlag).

Camagni, R. (2001) The economic role and spatial contradictions of global city-regions: The functional, cognitive and evolutionary context, in: A. J. Scott (Ed.) *Global City-Regions: Trends, Theory, Policy*, pp. 96–118 (Oxford: Oxford University Press).

Camagni, R. & Capello, R. (2004) The city network paradigm: Theory and empirical evidence, in: R. Capello & P. Nijkamp (Eds) *Urban Dynamics and Growth: Advances in Urban Economics*, pp. 495–532 (Amsterdam: Elsevier).

Camagni, R. & Pompili, T. (1991) La rendita fondiaria come indicatore della dinamica urbana: un'indagine empirica sul caso italiano, in: F. Boscacci & G. Gorla (Eds) *Economie locali in ambiente competitivo*, pp. 41–66 (Milan: Franco Angeli).

Camagni, R., Diappi, L. & Leonardi, G. (1986) Urban growth and decline in a hierarchical system: A supply-oriented dynamic approach, *Regional Science and Urban Economics*, 16(1), pp. 145–160.

Camagni, R., Capello, R. & Caragliu, A. (2013) One or infinite optimal city sizes? In search of an equilibrium size for cities, *The Annals of Regional Science*, 51(2), pp. 309–341.

Capello, R. (2000) The city network paradigm: Measuring urban network externalities, *Urban Studies*, 37(11), pp. 1925–1945.

Capello, R. (2002) Urban rent and urban dynamics: The determinants of urban development in Italy, *The Annals of Regional Science*, 36(4), pp. 593–611.

Capello, R. & Rietveld, P. (1998) The concept of network synergy in economic theory: Policy implications, in: K. Button P. Nijkamp & H. Priemus (Eds) *Transport Networks in Europe*, pp. 57–83 (Cheltenham: Edward Elgar).

Catin, M. (1991) Économies d'Agglomération et Gains de Productivité, *Revue d'Économie Régionale et Urbaine*, 5, pp. 565–598.

Chesnais, F. (1988) *Technical Co-operation Agreements Between Firms*, STI Review, Vol. 4 (Paris: OECD).

Conti, S. & Dematteis, G. (1995) Enterprises, systems and network dynamics: The challenge of complexity, in: S. Conti, E. Malecki & P. Oinas (Eds) *The Industrial Enterprise and Its Environment: Spatial Perspectives, Avebury, Aldershot*, pp. 217–242 (London: Sage).

Dijkstra, L. (2009) Metropolitan regions in the EU, *Regional Focus*, 1/2009, pp. 1–9.

Dijkstra, L., Garcilazo, E. & McCann, P. (2013) The economic performance of European cities and city regions: Myths and realities, *European Planning Studies*, 21(3), pp. 334–354.

Glaeser, E. L. (2008) *Cities, Agglomeration and Spatial Equilibrium* (Oxford: Oxford University Press).

Glaeser, E. L. (2011) *Triumph of the City: How Our Greatest Invention Makes Us Richer, Smarter, Greener, Healthier, and Happier* (New York: Penguin Books).

Glaeser, E. L., Kallal, H., Scheinkman, J. A. & Shleifer, A. (1992) Growth in cities, *Journal of Political Economy*, 100(6), pp. 1126–1152.

Henderson, J. (1974) The sizes and types of cities, *The American Economic Review*, 64(4), pp. 640–656.

Henderson, J. (1985) *Economic Theory and the Cities* (Orlando, FL: Academic Press).

Henderson, J. (1996) Ways to think about urban concentration: Neoclassical urban systems vs. the new economic geography, *International Regional Science Review*, 19(1&2), pp. 31–36.

Henderson, J. (2010) Cities and development, *Journal of Regional Science*, 50(1), pp. 515–540.

Krugman, P. (1991) *Geography and Trade* (Cambridge, MA: Cambridge University Press).

Marelli, E. (1981) Optimal city size, the productivity of cities and urban production functions, *Sistemi Urbani*, 1/2, pp. 149–163.

McCann, Ph. & Acs, Z. J. (2011) Globalization: Countries, Cities and Multinationals, *Regional Studies*, 45(1), pp. 17–32.

Meijers, E. J. & Burger, M. J. (2010) Spatial structure and productivity in US metropolitan areas, *Environment and Planning A*, 42(6), pp. 1383–1402.

Nijkamp, P. & Kourtit, K. (2011) New urban Europe: A place 4 all, *Urban Governance in the EU. Current Challenges and Future Prospects*, pp. 61–76 (Brussels: Committee of the Regions, European Commission, DG RTD).

Nijkamp, P. & Kourtit, K. (2012) The new urban Europe: Global challenges and local responses in the urban century, *European Planning Studies*, 21(3), pp. 291–315.

Parkinson, M., Meegan, R. & Karecha, J. (2014) City size and economic performance: Is bigger better, small more beautiful or middling marvellous? *European Planning Studies*, http://dx.doi.org/10.1080/09654313.2014.904998

Parr, J. B. (2002) Agglomeration economies: Ambiguities and confusions, *Environment and Planning A*, 34(4), pp. 717–731.

Phelps, N. A., Fallon, R. J., & Williams, C. L. (2001) Small firms, borrowed size and the urban-rural shift, *Regional Studies*, 35(7), pp. 613–624.

Richardson, H. (1972) Optimality in city size, systems of cities and urban policy: A sceptic's view, *Urban Studies*, 9(1), pp. 29–47.

Rosenthal, S. S. & Strange, W. C. (2004) Evidence on the nature and sources of agglomeration economies, in: J. Vernon Henderson & Jacques-Francois Thisse (Eds.), *Handbook of Regional and Urban Economics, Vol. 4 Cities and Geography*, pp. 2119–2171 (Amsterdam: Elsevier).

Rousseau M.-P. (1995) Y a-t-il une surproductivité de l'Île de France? in: M. Savy & P. Veltz (Eds) *Économie globale et Réinvention du Local*, pp. 157–167 (Paris: DATAR/éditions de l'aube).

Rousseau, M.-P. & Proud'homme, R. (1992) *Les avantages de la concentration parisienne* (Paris, FR: L'OEIL-IAURIF).

Sassen, S. (Ed.) (2002) *Global Networks, Linked Cities* (New York: Routledge).

Scott, A. J. (Ed.) (2001) *Global City-Regions: Trends, Theory, Policy* (Oxford: Oxford University Press).

Segal, D. (1976) Are there returns to scale in city size? *Review of Economics and Statistics*, 58(3), pp. 339–350.

Sveikauskas, L., Gowdy, J. & Funk, M. (1988) Urban productivity: City size or industry size, *Journal of Regional Science*, 28(2), pp. 185–202.

Veneri, P. & Burgalassi, D. (2012) Questioning polycentric development and its effects. Issues of definition and measurement for the Italian NUTS-2 regions, *European Planning Studies*, 20(6), pp. 1017–1037.

Williamson, O. E. (1985) *The Economic Institution of Capitalism* (New York: The Free Press).

Appendix. Details on the database

The database for the empirical analyses in this paper comprises the following 100 metro areas:

Århus, Aalborg, Aberdeen, Antwerpen, Arnhem, Augsburg, Bari, Belfast, Bilbao, Bologna, Bordeaux, Bremen, Bristol, Brno, Bydgoszcz, Caen, Cagliari, Cardiff, Charleroi, Clermont-Ferrand, Cluj-Napoca, Craiova, Czestochowa, Dijon, Dresden, Edinburgh, Eindhoven, Enschede, Erfurt, Exeter, Firenze, Freiburg im Breisgau, Gdansk, Genova, Gent, Graz, Groningen, Göteborg, Göttingen, Halle an der Saale, Hannover, Karlsruhe, Kiel, Kielce, Kingston-upon-Hull, Koblenz, Kosice, Kraków, Palmas/Sta. Cruz de Tenerife, Lefkosia, Leicester, Lille, Linz, Liverpool, Liège, Luxembourg, Lódz, Malmö, Maribor, Montpellier, Murcia, Nancy/Metz, Nantes, Napoli, Newcastle upon Tyne, Nottingham, Nürnberg, Ostrava, Oviedo, Palermo, Palma de Mallorca, Pamplona/Iruña, Plovdiv, Plzen, Porto, Portsmouth, Poznan, Regensburg, Rennes, Riga, Rostock, Santander, Sevilla, Sheffield, Stoke-on-Trent, Strasbourg, Szczecin, Tallinn, Tampere, Thessaloniki, Timisoara, Toulouse, Turku, Valencia, Valletta, Varna, Venezia, Vilnius, Wroclaw, Zaragoza.

Data encompass two observations per city, with the time coverage described in Section 4.

Borrowed Size, Agglomeration Shadows and Cultural Amenities in North-West Europe

MARTIJN J. BURGER*, EVERT J. MEIJERS**, MARLOES
M. HOOGERBRUGGE** & JAUME MASIP TRESSERRA**,†

*Department of Applied Economics, Tinbergen Institute, Erasmus University Rotterdam, Rotterdam, The Netherlands, **Faculty of Architecture and the Built Environment, Delft University of Technology, Delft, The Netherlands, †Centre of Land Policy and Valuations, Polytechnic University of Catalonia, Barcelona, Spain

ABSTRACT *It has been argued that the concept of "borrowed size" is essential to understanding urban patterns and dynamics in North-West Europe. This paper conceptualizes this idea and provides an empirical exploration of it. A place borrows size when it hosts more urban functions than its own size could normally support. A borrowed size for one place means that other places face an "agglomeration shadow" because they host fewer urban functions than they would normally support. This paper explores the extent to which size and function are related for places in North-West Europe and tries to explain why one place borrows size while the other faces an agglomeration shadow by examining the position of places within the regional urban system. The presence of urban functions was approximated using high-end cultural amenities. We conclude that the largest places in their functional urban area (FUA) are better able to exploit their own mass. The largest place in a FUA is also better able to borrow size from nearby places and from (inter)national urban networks than the lower-ranked places.*

1. Introduction

1.1. *Rise of the Consumer City*

In the past, cities were primarily seen as centres of production, while nowadays they are increasingly perceived as centres of consumption (Glaeser *et al.*, 2001; Rappaport, 2008). Although nominal wages increase with the city size, the costs of living increase even more, indicating that people are willing to give up real wages to enjoy the consumer amenities that are typically found in large cities (Tabuchi & Yoshida, 2000). This trade-off holds especially for the higher-educated part of the workforce (Lee, 2010; Dalmazzo & De Blasio, 2011).[1] Amenities, here, are not only the aesthetic properties of large cities such as historical buildings and heritage sites (e.g. Van Duijn & Rouwendal, 2013), but also the presence of a variety of specialized goods and services (e.g. Berry & Waldfogel, 2010; Burger et al., 2013) and the provision of specialized public services (e.g. Dahlberg et al., 2012; Ebertz, 2013). Traditionally, universities, hospitals, stadiums, concert halls, theatres, galleries, museums and higher-order retail functions are present in large cities, while inhabitants of towns, villages and hamlets do not have ready local access to these amenities. The presence of consumer amenities is also increasingly important in the location decisions of firms and households, which is reflected in the higher growth rates of amenity-rich places (Glaeser et al., 2001; Clark et al., 2002; Chen & Rosenthal, 2008; Markusen & Schrock, 2009) as well as in the recent increase in exchange commuting in many Western societies,[2] that is, people living in the amenity-rich central cities and working in the suburban areas (Van der Laan, 1998; Aguilera, 2005; De Goei et al., 2010; Burger et al., 2011).

Consumer amenities are known to be strongly dependent on the size of the local population (Berry & Parr, 1988). In particular, the presence of a large quantity of high-end consumer functions is strongly associated with the size of a city and, as such, a manifestation of the presence of urbanization economies through consumption (Glaeser et al., 2001; Rosenthal & Strange, 2004).

1.2. *From City to Urban Network*

Parallel to the rise of the consumer city, we see changes in the structure of urban systems and an increasing geographical scope for social and economic processes (see, e.g. Camagni, 1993; Batten, 1995; Champion, 2001). At the metropolitan scale, suburban areas are increasingly emerging into local centres that develop their own economic activities and, in some cases, start competing with the original urban centre. At higher spatial scales, there is an increasingly complex formation of functional linkages between historically separated urban areas. These developments are reflected in the increasing popularity of concepts such as polycentric urban regions (Kloosterman & Musterd, 2001; Parr, 2004) and mega-city regions (Hall & Pain, 2006; Florida et al., 2008; Hoyler et al., 2008). Such urban networks are characterized by a considerable degree of spatial integration between different places and, in many cases, complementary relationships between places (Champion, 2001; Parr, 2004; Meijers, 2007a). Accordingly, cities and towns within an urban network are thought to have different sectoral and functional specializations (Meijers, 2008; Van Oort et al., 2010), where some places appear to specialize in delivering high levels of consumer amenities signified by the sometimes considerable amenity gaps between places within a region (BBSR, 2011; Silver, 2012). This study analyses how network economies allow cities to borrow size from other places at different geographical scales.[3]

1.3. *Borrowed Size and Agglomeration Shadows*

Within networked constellations, there is not necessarily a relationship between the size of a place and the functions it fulfils (Batten, 1995; Meijers, 2007b) because the critical mass to support particular functions can be obtained from the wider urban network. In other words, the "impact zone" (cf. Clark, 2004) of urban functions increases.[4] Due to the presence of spatial interdependencies, smaller places can "borrow size" and host functions that they could not have hosted in isolation. The concept of borrowed size was introduced by Alonso (1973) to explain this disconnection between the size and function of smaller cities that were part of a megalopolitan urban complex:

> [t]he concept of a system of cities has many facets, but one of particular interest ... is the concept of borrowed size, whereby a small city or metropolitan area exhibits some of the characteristics of a larger one if it is near other population concentrations. (Alonso, 1973, p. 200)

More precisely, he suggests that smaller places may "borrow" some of the agglomeration benefits of their larger neighbours, while avoiding the agglomeration costs. Even back in the 1970s, Alonso, himself hinting at the pattern of urbanization in the North-Eastern USA, noted that borrowed size was a key issue to understanding the urbanization pattern of North-West Europe: processes of borrowed size are

> also quite visible ... in certain European urban patterns, such as those of Germany and the Low Countries, whose cities, quite small by our standards, apparently achieve sufficient scale for the functioning of a modern economy by borrowing size from one another. This phenomenon transforms the issue of the size and growth of a city by redefining it to include, in some degree, its neighbours.
>
> (Alonso, 1973, p. 200)

Despite the promise that the concept of borrowed size explains the pattern of urbanization in these countries, it has received only limited attention in economics and geography (Phelps & Ozawa, 2003).

Although smaller places can potentially host urban functions that are normally only found in larger cities through this process of borrowing size, increased spatial competition between nearby places can also lead to a situation in which a place hosts fewer functions than it could normally support. Some have conceptualized this situation as being an "agglomeration shadow" (Fujita et al., 1999; Dobkins & Ioannides, 2001; Partridge et al., 2009). Building on Central Place Theory (Christaller, 1933) and urban systems theory (Berry, 1964), New Economic Geography predicts a shadow effect from large cities over their surroundings, which means that the growth of areas near higher-order places will be limited due to competition effects. In terms of the presence of consumer amenities, it can be expected that places in proximity to large cities actually have fewer consumer amenities than isolated places of similar size. This expectation reflects that the provision of higher-order amenities in larger cities is often facilitated by control over the wider region as a trade area for high-end functions, where subordinate centres help to provide the minimum demand threshold to support these functions (Wensley & Stabler, 1998; Thilmany et al., 2005). In this context, the overrepresentation of high-end amenities in the largest cities is driven by the advantages of locating closest to the largest customer base. Although competitors try to avoid each other, the agglomeration of amenities in large cities can be explained by savings on the travel, search and transaction costs associated with multi-purpose trips (Eaton & Lipsey, 1982; Glaeser et al., 2001). Obviously, the benefits from the co-location of amenities for consumers imply that agglomeration is to their advantage as well. Anticipating consumer behaviour, amenities that share their location with other competing and complementary amenities are more likely to attract consumers than an otherwise identical amenity located on its own. Accordingly, the agglomeration of amenities creates advantages for both the consumers and the producers of these services (Mulligan, 1984).

1.4. *Research Question*

In sum, increasing spatial interdependencies between places can result in a disconnection between the size of a place and its function, here measured by the presence of particular consumer amenities. On the one hand, minimum threshold demands for consumer amenities can be supported by the population of places in the urban network, which provides additional external demand. On the other hand, the presence of amenities elsewhere in the network results in external competing sources of supply drawing (potential) local consumers away from these areas (Mushinski & Weiler, 2002). However, it remains unclear which places profit and which places lose in light of increasing spatial interdependencies. In other words, which places borrow size and which places face agglomeration shadows?

In this paper, we narrow this question down to a more specific one addressing the capacity of places to borrow size. Focusing on high-end amenities in urban Europe related to cultural venues (museums, operas, theatres and public art institutions) and cultural events (music events, film festivals and art exhibitions), we examine the extent to which the population size of a place and its position in the wider regional, national and international networks affects the presence of these amenities. In particular, we are interested in how the position of a place in the regional urban network affects the relationship between the population variables and the cultural amenities present in that place. Moreover, we analyse the ability of places to borrow size by analysing national and international accessibility. The expectation is that places that are relatively well (internationally) connected are able to borrow size from other places in Europe or even beyond. So, this paper explores the question whether the position of a place in the regional urban system affects its capacity to borrow size locally, regionally and (inter)nationally.

The remainder of this paper is organized as follows. The next section presents the research approach, including concepts, data and variables, and the modelling approach. The empirical results are presented in Section 3. We conclude with a discussion of our findings in Section 4.

2. Research Approach

Cultural amenities provide a good case study for examining the occurrence of borrowed size and agglomeration shadows because their presence is generally strongly related to the size of a place. A disconnection between the size and function of a place is an indication of either borrowed size (if a place has more cultural amenities than expected given its size) or agglomeration shadows (if a place hosts fewer cultural amenities than expected given its size). We consider these concepts to be two sides of the same coin, and consequently the way to operationalize them is similar. This relationship also means that a borrowed size effect for one place implies that one or more other places are in an agglomeration shadow.

While we believe that the processes of borrowed size and agglomeration shadows can occur on many territorial scales, it would appear logical to assume that they are most often manifested on the scale of functional urban regions because nearby cities are more likely to borrow size from each other due to their proximity. At the same time, good accessibility to places beyond the own metropolitan region may also foster processes of borrowed size or agglomeration shadows on a national or international scale. For instance, the cluster of theatres in places like New York or London builds on a customer base that is perhaps even global in scope and might limit the support base for cultural amenities in the cities from which these customers come. This observation is another reason to focus on high-end cultural amenities, because consumers are willing to travel longer distances for these types of amenities. As accessibility strongly determines the ability of places to borrow size from each other, it is taken into account in this study.

2.1. *Data and Variables*

To analyse the moderating effect of network position on the relationship between the size of a place and the amenities that it provides, we use data from the German Federal Institute for Research on Building, Urban Affairs and Spatial Development (BBSR, 2011). This database provides information on the presence of metropolitan functions in places (mostly defined at LAU-2) in the urbanized part of Europe. In this study, we focus on the presence of cultural events and venues in 2794 West-European cities, towns and villages for the years 2006–2007. The population size of the places in the database ranges from 61 (Gerstengrund) to almost 3.5 million (Berlin). Most LAU-2 spatial units are located in France (35%) and Germany (29%), followed by Switzerland (14%), the Netherlands (8%), Austria (7%), Belgium (6%), and Luxembourg (1%).

To capture the presence of cultural amenities in a single indicator, we developed an index of cultural amenities that is based on six indicators: number of theatres, number of opera houses and music theatres, number of large music events hosted,[5] number and quality of public art institutions,[6] number of art fairs and film festivals, and number of art galleries. The place with the highest average score on the six dimensions receives a score of 1, while other cities receive a score relative to the highest ranked city, where the minimum score is 0. A more detailed overview of the measurement of the various indicators is provided by BBSR (2011).

Figure 1 shows the distribution of cultural amenities across European places. As can be expected, this figure shows that cultural amenities are strongly concentrated in the largest cities in North-West Europe (Berlin, Paris and Vienna; see Table 1). At the same time, most places (78%) lack high-end cultural amenities, in that they score zero on the Cultural Amenities Index, which is also signified by the low average presence of these amenities (0.0047) across all places in North-West Europe. Places with fewer than 50,000 inhabitants score, on average, 0.001 on the Cultural Amenities Index, while places with 100,000–350,000 inhabitants and places with more than 350,000 inhabitants (the 25 largest places in North-West Europe) score 0.029 and 0.257, respectively.

In this research, we examine the extent to which the presence of high-end cultural amenities in a place is first of all dependent on the local population size (LAU-2). Second, we take into account population variables that represent (1) the size of the population in the rest of the functional urban area (FUA) in which the place is situated and (2) the size of the population that can be accessed on a national and international scale (outside of the own

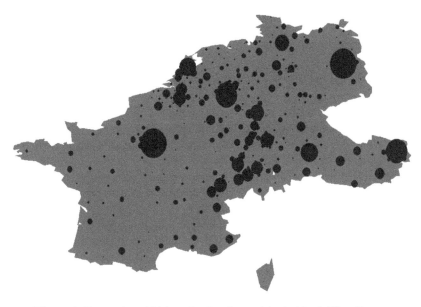

Figure 1. Geography of high-end cultural amenities in North-West Europe.
Source: Own calculations based on BBSR (2011).

Table 1.   Scores of 25 largest places in North-West Europe on Cultural Amenities Index

Place (LAU-2)	Cultural Amenities Index	Population (millions)
Berlin	1.00	3.46
Paris	0.96	2.23
Vienna	0.65	1.70
Cologne	0.64	1.01
Munich	0.42	1.35
Amsterdam	0.36	0.77
Hamburg	0.29	1.79
Frankfurt am Main	0.29	0.68
Stuttgart	0.25	0.61
Bruxelles	0.23	1.09
Antwerp	0.17	0.48
Zurich	0.17	0.37
Düsseldorf	0.17	0.59
Hannover	0.12	0.52
Nürnberg	0.11	0.51
Lyon	0.10	0.48
Essen	0.09	0.57
The Hague	0.08	0.49
Marseille	0.06	0.85
Dortmund	0.06	0.58
Bremen	0.06	0.66
Rotterdam	0.06	0.59
Toulouse	0.06	0.44
Bochum	0.02	0.37
Duisburg	0.01	0.49

FUA). The question then is whether or not a place profits from being located in a larger metropolitan area or having access to a large population (inter)nationally.

Here, our definition of FUAs follows that of the ESPON 1.4.3 project (Peeters, 2011), where FUAs represent labour market regions that encompass cities, towns and villages that are economically and socially highly interdependent. The types of FUAs range from mega-city regions, such as Paris (11.1 million habitants), Milano (6.0 million inhabitants), and Dortmund (5.3 million inhabitants), to self-contained villages in more rural areas with just over 10,000 inhabitants. The average size of a place in our database is just over 30,000 inhabitants, while the average size of a place that is the largest in their FUA is almost 130,000. Overall, the places in our database are nested in 417 FUAs, meaning that each FUA consists, on average, of 6–7 LAU-2 units. By analysing the places within FUAs, local proximity is taken into account because these FUAs represent labour market regions with a maximum radius of 60 kilometres.

To capture national and international accessibility, we make use of the absolute potential accessibility to population by road, rail and air (multimodal) from the ESPON research reported in Spiekermann and Wegener (2006). This accessibility measure takes into account the population outside of the FUA weighted by the travel time to go there.[7] The most accessible regions can be found in the mega-regions with major international airports such as the Rhine-Main (Frankfurt am Main) and Randstad (Amsterdam); the least accessible regions can be found in peripheral France, Germany and Austria.

As additional control variables, we introduce a dummy variable indicating whether a place contains a UNICEF heritage site (as an indicator of the historical presence of amenities), GDP per capita (measured at the FUA level), and country dummies. These country dummies capture, among others, institutional differences between countries with regard to the production of public goods and services, such as cultural venues. Descriptive statistics are presented in Table 2; all variables are measured for the year 2006, unless indicated otherwise.

Where a positive effect of the population on the rest of the FUA and national and international accessibility on the presence of high-end cultural amenities would indicate the presence of borrowed size effects and urban network economies, a negative effect would indicate the presence of agglomeration shadows or, in other words, spatial competition effects. Yet, we do not expect these variables to be significant. After all, one place may perhaps borrow size from its wider region or (inter)nationally, but another place may

Table 2. Descriptive statistics of variables included in the analysis ($N = 2794$)

	Mean	Standard deviation	Minimum	Maximum
Cultural Amenities Index	0.0047	0.037	0	1
Local population (100 k)	0.32	1.13	0.00061	34.61
Population rest FUA (mln)	2.16	3.07	0	11.17
(Inter)national accessibility (10 mln)	6.85	1.51	2.96	10.01
GDP per capita FUA (k)	32.39	8.47	13.00	54.00
Dummy heritage site	0.02	0.15	0	1
Dummy not largest place in FUA	0.85	0.35	0	1
• Dummy second largest place in FUA	0.11	0.32	0	1
• Dummy third largest place in FUA	0.08	0.29	0	1
• Dummy fourth largest place in FUA	0.07	0.25	0	1
• Dummy other not largest places in FUA	0.58	0.49	0	1

experience an agglomeration shadow effect. We assumed that whether a place borrows size or faces such agglomeration shadows depends on its position in the wider urban network (or metropolitan urban system). Therefore, we examine the extent to which the rank of a place within the FUA (affects and) moderates the relationship between the population variables and the Cultural Amenities Index. More specifically, we test the extent to which local population size, the population in the rest of the FUA and national and international accessibility can successfully be exploited by including an interaction term between the population variables and the rank of a place in the FUA in which it is located. Here, we assume that second or lower rank cities have more difficulties borrowing size compared to the largest city in an FUA. This assumption builds on the observation that co-location (or agglomeration) is beneficial for both retailers and consumers.

2.2. *Empirical Model*

Our dependent variable, the Cultural Amenities Index, can be perceived as a bounded variable that takes a minimum value of 0 (no amenities) and a maximum value of 1 (city with the highest score on the Amenities Index). Although indices (which can likewise be perceived as proportions) are often treated as if they were continuous, an estimation by ordinary least squares in a linear regression framework is likely to result in inefficient and biased estimates for the parameters. The most common way to address proportions data is the fractional logit model (Papke & Woolridge, 1996, 2008). However, this model often suffers from overdispersion if the response variable is non-binomially distributed in that the probability masses are concentrated at the boundaries. In the case of our data, many places do not have any high-end cultural amenities (in that they score zero), while only a few places score very high (see Figure 2). To account for this characteristic, we make use of a beta regression model (Kieschnick & McCullough, 2003; Ferrari & Cribari-Neto, 2004). This type of model utilizes the beta distribution, which makes it appropriate for modelling both binomially and non-binomially distributed response variables. At the same time, the interpretation of the results is similar to that for a logistic regression (Schmid *et al.*, 2013). In addition, beta regression models provide a natural way to address overdispersion by including an additional parameter (phi) to adjust the conditional variance of the proportion outcome. Because our proportions data contains zeros (no high-end amenities) and ones (highest ranked place), we make use of inflated beta regression models. Zero-or-one inflated beta regression involves three parts: a logistic regression model for whether or not the proportion equals 0, a logistic regression model for whether or not the proportion equals 1, and a beta regression model for the proportions between 0 and 1. Because our data particularly suffers from probability masses concentrated at 0, we predominantly focus on the zero-inflated part, herewith also acknowledging that there may be population thresholds for providing certain amenities. A more elaborate discussion of these models can be found in Cook *et al.* (2008) and Ospina and Ferrari (2012).

3. Results

3.1. *Empirical Results*

Table 3 reports the coefficients of the inflated beta regressions on our Cultural Amenities Index. Because we are not only interested in the statistical significance of effects but also

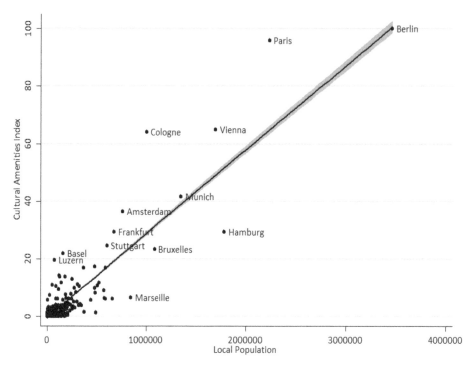

Figure 2. Relationship between local population size and cultural amenities.

in the substantive and practical significance, we also report the average marginal effects (AMEs). With regard to our population variable, the reported AMEs refer to a change in the high-end Cultural Amenities Index as a result of an assumed increase in the local population by 100,000, a change in the population living in the rest of the FUA by 1 million, and a change in (inter)national accessibility by 10 million people. Likewise, the AME for GDP per capita refers to a change in the high-end Cultural Amenities Index as a result of an assumed increase in GDP per capita by 1000 euro, while for the dummy variables, the AME refers to a change in the high-end cultural amenities variable as a result of a change in the dummy variable from 0 to 1.

Focusing on our baseline model (Table 3, Column 1), we find no effect from GDP per capita,[8] but a positive effect from the cultural heritage site dummy on the presence of high-end cultural amenities in a place. Turning to our primary variables of interest, the results indicate that local population size is the most important determinant for the presence of high-end amenities in a place, while the population in the rest of the FUA and (inter)-national accessibility is found to be less important, which is consistent with our expectations. The AME of the local population size is 0.007, meaning that an increase in population size by 100,000 increases the high-end amenities (an index with minimum 0 and maximum 1), on average, by 0.007.

The AME is low, which signifies that there is a considerable population threshold that must be met to host high-end cultural amenities in the first place. Figure 3 shows how the AME varies across the size of the place for places between 10,000 and 1,000,000 (all observations that fall within three standard deviations of the mean, excluding outliers).

Table 3. Zero-and-one inflated beta regression on Cultural Amenities Index—regression coefficients and AMEs.

	Model 1		Model 2		Model 3	
	Coefficients	AME	Coefficients	AME	Coefficients	AME
Proportion part						
Local population (100 k)	0.297 (.045)**	0.0071 (.0006)**	0.248 (.036)**	0.0057 (.0005)**	0.253 (.036)**	0.0056 (.0005)**
Population rest FUA (mln)	−0.060 (.021)**	−0.0002 (.0001)	−0.018 (.021)	−0.0000 (.0001)	−0.000 (.022)	0.0001 (.0001)
(Inter)national accessibility (10 mln)	0.044 (.021)	−0.0001 (.0001)	0.079 (.030)**	0.0001 (.0001)	0.084 (.031)**	0.0001 (.0001)
GDP per capita FUA (k)	0.008 (.006)	0.0000 (.0000)	0.019 (.008)**	0.0001 (.0000)*	0.022 (.008)**	0.0001 (.0000)**
Dummy heritage site	0.769 (.203)**	0.0037 (.0009)**	0.597 (.190)**	0.0029 (.0009)**	0.603 (.189)**	0.0029 (.0008)**
Dummy not largest location in FUA			−0.748 (.085)**	−0.0030 (.0004)**		
• Second largest location in FUA					−0.605 (.089)**	−0.0022 (.0004)**
• Third largest location in FUA					−0.739 (.108)**	−0.0031 (.0005)**
• Fourth largest location in FUA					−0.742 (.109)**	−0.0032 (.0005)**
• Other not largest locations in FUA					−0.927 (.110)**	−0.0040 (.0005)**
Zero-inflated part						
Local population (100 k)	−4.781 (.291)**		−4.579 (.352)**		−4.258 (.358)**	
Population rest FUA (mln)	−0.032 (.045)		−0.039 (.045)		−0.06 (.045)	
(Inter)national accessibility (10 mln)	0.226 (.059)**		0.214 (.061)**		0.175 (.063)**	
GDP per capita FUA (k)	0.010 (.012)		0.008 (.012)		0.002 (.012)	
Dummy heritage site	−0.702 (.398)		−0.658 (.399)		−0.693 (.401)	
Dummy not largest location in FUA			0.227 (.207)			
• Second largest location in FUA					−0.048 (.224)	
• Third largest location in FUA					0.341 (.243)	
• Fourth largest location in FUA					0.520 (.283)	
• Other not largest locations in FUA					0.574 (.243)*	
Number of Observations	2794		2794		2794	
Wald chi-square	246.0**		521.0**		548.3**	
ln phi	3.86 (.129)**		4.06 (.135)**		4.08 (.134)**	
Country dummies +	Yes		Yes		Yes	

Notes: Robust standard errors in parentheses. All models are estimated with intercept. The one-inflated part is not displayed as it is an intercept-only model. For the dummy variables the AME refers to a change of the variable from 0 to 1. AME, average marginal effect.

*p < .05.
**p < .01.

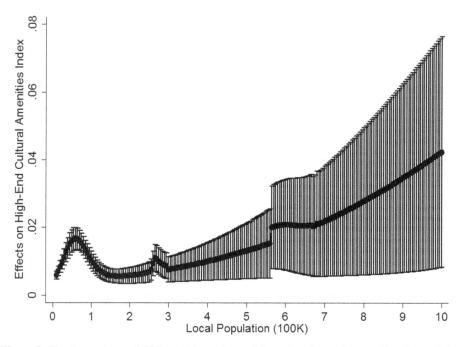

Figure 3. Fitted equation and 95% confidence interval for cultural amenities and local population size (AME).

Although an increase in population size has only a small effect on the presence of cultural amenities in places with fewer than 100,000–200,000 inhabitants, the AME of local population size on the proportion of amenities steadily increases after this threshold is reached. To exemplify, we find that for a place of 150,000–200,000 inhabitants, a 100,000 increase in the local population size would increase the Cultural Amenities Index by approximately 0.006. On the contrary, for a city of 500,000 inhabitants, 750,000 inhabitants, and 1,000,000 inhabitants, a similar increase would result in an increase in the Cultural Amenities Index by 0.013, 0.025 and 0.042, respectively.[9] A spike in the graph can also be observed for places having between 10,000 and 100,000 inhabitants, indicating that the high-end Cultural Amenities Index includes some amenities for which the population threshold is limited. This result is consistent with earlier work by Camagni *et al.* (1986, 2014), who indicate that when the population in a place increases and moves towards the maximum size compatible with its rank, this place becomes a potentially suitable location for higher-order urban functions, a result of achieving a critical demand size for these functions.

Consistent with our expectations, the AME from the size of the rest of the FUA and (inter)national accessibility on the score for the high-end Cultural Amenities Index is insignificant in our baseline specification. However, the position of a place in the FUA also has an effect on the presence of high-end cultural amenities (see Models 2 and 3 in Table 3). Not being the largest city decreases the cultural amenities by approximately 0.003 on average, holding all other variables constant. Likewise, the lower the rank of a place in its FUA, the fewer high-end amenities are present in that place. Compared to being the largest city in the FUA and holding everything else constant, the second largest places score, on average, 0.002 lower on the Cultural Amenities Index, while the fifth or lower-ranked places score, on average, over 0.004 lower. Accordingly, it can be inferred that the largest places within a FUA fulfil a central place function with regard to hosting high-end cultural amenities and, hence, borrow size from elsewhere, thereby casting an agglomeration shadow over the surrounding places.

At the same time, the rank of a place in a FUA may also affect the extent to which it can exploit its local size as well as the extent to which it can profit from the size of the rest of the FUA and from national and international accessibility. To test whether this situation is indeed the case, we estimated additional regressions, including interaction terms between the population variables and the dummy variable indicating whether a place is the largest within an FUA.[10] Because interaction effects are generally difficult to interpret in non-linear models, we graphically show how the effect of the position in the FUA on the presence of high-end cultural amenities depends on local population size, population in the rest of the FUA, and national and international accessibility. The regression table can be found in Appendix 1.

Figures 4–6 graphically show how the position in the FUA affects the capacity to exploit local potential (Figure 4) or the potential to borrow size (exploit) from the critical mass present in the wider FUA (Figure 5) or (inter)nationally (Figure 6). Several conclusions can be drawn from these graphs. First, places are better able to profit from being the largest in their FUA when they are small (Figure 4). Estimating the AMEs for being the largest place in the FUA for different place sizes,[11] we find that the AME of local population size on the Cultural Amenities Index is stronger for places that are the largest in their FUA. At the same time, we see that the positive moderating effect of

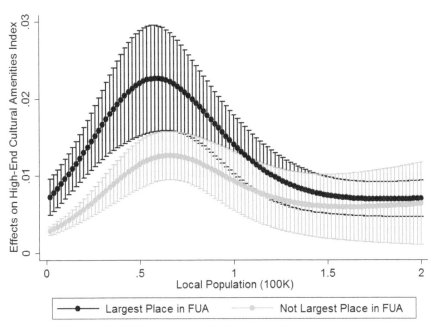

Figure 4. Exploiting local potential—AMEs.

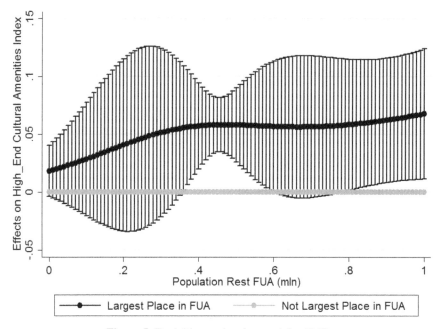

Figure 5. Exploiting regional potential—AMEs.

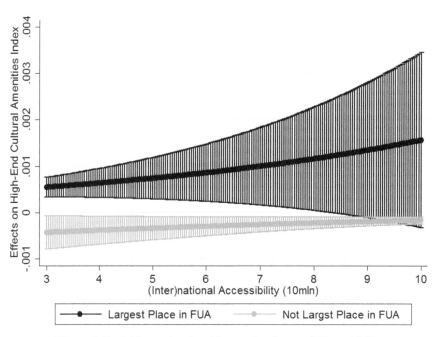

Figure 6. Exploiting national and international accessibility—AMEs.

being the largest place in an FUA on the relationship between local population size and the presence of high-end cultural amenities is the most pronounced for places with between 40,000 and 80,000 inhabitants.

Second, places tend to profit more from the size of their FUA when they are the largest in their FUA (Figure 5).[12] For the places that are the largest in their FUA, a 1 million increase in population in the rest of the FUA would result, on average, in an increase of 0.05 in the high-end Cultural Amenities Index. On the contrary, the size of the rest of the FUA does not affect the presence of high-end cultural amenities in other places in the FUA.

Third, (inter)national accessibility has a much more positive impact on the presence of high-end cultural amenities in the largest places in the FUAs (Figure 6). From this result, it can be concluded that the largest places in an FUA profit most from (inter)national accessibility and borrow size in (inter)national urban networks to support the production of high-end cultural amenities. At the same time, it should be noted that the significance of this finding is limited: an increase in (inter)national accessibility by 10 million would, even for the largest places in the FUAs, only result in an increase in the Cultural Amenities Index by 0.001–0.002. To exemplify, the most accessible places in North-West Europe would score, on average, 0.007–0.014 higher on the Cultural Amenities Index than the least accessible places, holding everything else constant.

4. Conclusion

It has been argued that the concept of "borrowed size" is essential for understanding urban patterns and dynamics in North-West Europe. This paper is one of the first to empirically explore the concept of "borrowed size". It is likely that a place borrows size when it hosts more urban functions than its own size could normally support. The opposite, however, also occurs: a place can host fewer urban functions than it should normally be capable of supporting. Often, this situation is due to competition from nearby places or from places with good (inter)national accessibility. To capture this phenomenon, we proposed using the concept of "agglomeration shadows" and presented it as the antonym of "borrowed size".

In this study, the presence of urban functions was measured using information on high-end cultural amenities such as theatres, operas, public art institutions, concerts, and art and film festivals. We examined the extent to which local population size, on the one hand, and population potential elsewhere in the FUA or even beyond, on the other hand, affects the presence of these amenities. The empirical results indicate that, as expected, the size of a place is the most important factor explaining the presence of cultural amenities. Whether a place is located in a small or large FUA or has high (inter)national accessibility, giving access to a large (international) population potential, mattered less. This result indicates that the geographical scope of consumption is still very local. However, the presence of high-end cultural amenities in a place is also dependent on the position of that place in its FUA. To capture this position, we used measures that refer to the rank of a city within its functional urban region. The results show that cultural amenities are disproportionally concentrated in the places that are the largest in their FUA and that places profit more from being the largest in their FUA when they are small in terms of population size. In addition, places profit more from the size of the rest of the FUA and (inter)national accessibility when they are the largest place within their FUA.

In sum, the distribution of high-end cultural amenities in North-West Europe still follows a Christallerian logic, signified by the concentration of these consumer services within the largest place in a FUA as a result of the higher level of local demand and the accessibility of these often central places to clients from neighbouring places. At the same time, the places that lie in the shadow of these central places are confined to lower levels of high-end cultural amenities than could be expected given their size. In other words, on average, larger cities cast a shadow over smaller neighbouring cities (as predicted by the New Economic Geography) rather than these smaller cities borrowing size from their larger neighbour (as suggested by Alonso). This study indicates that the largest cities in a FUA profit more from the size of the rest of the area in which they are located and from having access to a large (inter)national population. The lower order places, which are not the largest city in the FUA, are less likely to borrow size from other places because they are more likely to lie in the shadow of a first-rank city and face spatial competition effects. Accordingly, large cities profit most from borrowed size effects, whereas small- and medium-sized places in Europe typically flourish in iso-lation.

The strong presence of agglomeration economies in consumption questions the extent to which small- and medium-sized cities can compete with large cities on the basis of cultural amenities. All the same, a caveat of this research is that we have predominantly examined high-end cultural amenities; small- and medium-sized cities near large urban agglomera-tions do not necessarily have to be confined to lower-end cultural amenities because these economic activities can follow a different location pattern. Although Madonna, Radio-head, Lady Gaga, U2, and Justin Bieber might avoid small- and medium-sized cities, more locally and regionally oriented musicians do not. Likewise, small- and medium-sized cities typically lack film and art festivals, but can host smaller-scale cultural events. Hence, future research should also try to incorporate other types of consumer ser-vices. We can extend this observation even further, as future research could focus on a much broader variety of urban functions beyond only high-end cultural amenities. It could well be that the complementarity of urban functions between places comprising a metropolitan region is the key to understanding processes of borrowed size and agglom-eration shadows. Perhaps a place faces an agglomeration shadow in one respect, but borrows size in another. Likewise, future research could focus on using more detailed measures for the position of places in areas and networks, also examining the extent to which being the largest place in the wider region or country affects the presence of urban functions. Moreover, it is interesting to analyse the multiple factors that determine the ability of some places to derive critical mass from other regions. Here, one can not only consider the quality of infrastructure and ease of access (see also Agnoletti *et al.*, 2014) but also other factors, such as the general quality of the services provided and the overall level of demand and supply.

Acknowledgements

We thank BBSR for sharing their data on metropolitan functions. We also benefitted from useful feedback from two anonymous referees and participants of the IRPET conference in Florence on "Welfare and competitiveness in the European polycentric urban structure: Which role for metropolitan, medium and small cities?" This research is part of the research project "Networks, Agglomeration and Polycentric Metropolitan Areas: New

Perspectives for Improved Economic Performance" within the Knowledge for Strong Cities (KKS) research programme.

Funding

This research was funded by the Dutch Science Council NWO and Platform31.

Notes

1. This viewpoint typically builds on the Rosen (1979) and Roback (1982) approach.
2. For a criticism of the amenity-growth nexus, see Peck (2005) and Storper and Scott (2009).
3. See also the study by Giffinger and Suitner (2014), which distinguishes between three different levels of integration of urban agglomeration, namely the micro, meso and macro levels. Micro-level polycentricity points at the level of internal integration of metropolitan regions, meso-level polycentricity defines dense national urban networks and macro-level polycentricity describes the embedding of metropolitan regions in wider transnational or global networks of metropolises.
4. This increased impact zone is consistent with observations that small- and medium-sized places have joined the urban competition for investments and labour (Smidt-Jensen *et al.*, 2009).
5. This indicator is based on the tour dates for bands and orchestras covering a broad range of musical styles: The Rolling Stones, Madonna, Sting and Bon Jovi (pop and rock); Anna Netrebko, Anne-Sophie Mutter, the Vienna Philharmonic and the New York Philharmonic Orchestra (classical music); and Cats (musical) as well as annual international jazz festivals.
6. Public art institutions include public art museums and galleries, non-profit or private art clubs, art foundations and associations, art archives and related libraries, and art colleges or universities teaching art, cinematic art, photography, design and graphic arts.
7. Although the multimodal accessibility measure was originally developed for NUTS-3 regions, it has been adapted to FUAs for this research.
8. Here, it must be mentioned that our measure of GDP per capita is calculated at the FUA level, whereas the Cultural Amenities Index is calculated at the place level, meaning that our GDP variable does not capture the, in some cases, strong heterogeneity in income between nearby places (e.g. centres and suburban places). However, income that is generated in one place can easily be spent in another place, and a measure of income at the FUA level is better able to control for this. When we control for the position of a place in a FUA, we find a small positive AME of GDP per capita, where an increase of GDP per capita by 1000 euro in the FUA results in an increase in the high-end Cultural Amenities Index by 0.0001.
9. The results of these estimations are available upon request.
10. Similar regressions were run for interactions between the rank order and the local population, the population in the rest of the FUA and national and international accessibility. For brevity, these graphs are not displayed. However, these estimations do not lead to different conclusions regarding borrowed size and spatial competition effects.
11. Because the size of the largest place that is not the largest in its FUA (Oberhausen, Germany) is just over 200,000, we do not show the AMEs for larger local population sizes in this graph.
12. Here, we focus on regional hinterlands with up to 1 million inhabitants because there are few places (less than 0.5%) that are the largest in their FUA for which the size of the rest of the FUA is more than 1 million inhabitants.

References

Agnoletti, C., Bocci, C., Lommi, S., Lattarulo, P. & Marinari, D. (2014) First and second tier cities in regional agglomeration models, *European Planning Studies*, doi:10.1080/09654313.2014.905006.

Aguilera, A. (2005) Growth in commuting distances in French polycentric metropolitan areas: Paris, Lyon and Marseille, *Urban Studies*, 42(9), pp. 1537–1547.

Alonso, W. (1973) Urban zero population growth, *Daedalus*, 102(4), pp. 191–206.

Batten, D. F. (1995) Network cities: Creative urban agglomerations for the 21st century, *Urban Studies*, 32(2), pp. 313–327.

BBSR (2011) *Metropolitan Regions in Europe. BBSR-online-publikation*, Nr. 01/2011, Bonn: Federal Institute for Research on Building, Urban Affairs and Spatial Development (BBSR) within the Federal Office for Building and Regional Planning (BBR), January.

Berry, B. J. L. (1964) Cities as systems within systems of cities, *Papers of the Regional Science Association*, 13(1), pp. 46–163.

Berry, S. & Waldfogel, J. (2010) Product quality and market size, *Journal of Industrial Economics*, 58(1), pp. 1–31.

Berry, B. J. L., Parr, J. B., Epstein, B. J., Ghosh, A. & Smith, R. H. T. (1988) *Market Centers as Retail Locations* (Englewood Cliffs, NJ: Prentice Hall).

Burger, M. J., De Goei, B., Van der Laan, L. & Huisman, F. M. J. (2011) Heterogeneous development of metropolitan spatial structure: Evidence from commuting patterns in English and Welsh city-regions, *Cities*, 28(2), pp. 160–170.

Burger, M. J., Meijers, E. J. & Van Oort, F. G. (2013) Regional spatial structure and retail amenities in the Netherlands, *Regional Studies*, doi:10.1080/00343404.2013.783693

Camagni, R. (1993) From city hierarchy to city network; reflections about an emerging paradigm, in: T. R. Lakshmanan & P. Nijkamp (Eds) *Structure and Change in the Space Economy: Festschrift in Honour of Martin Beckmann*, pp. 66–87 (Berlin: Springer).

Camagni, R., Capello, R. & Caragliu, A. (2014) The rise of second-rank cities: What role for agglomeration economies? *European Planning Studies*, doi:10.1080/09654313.2014.904999

Camagni, R., Diappi, L. & Leonardi, G. (1986) Urban growth and decline in a hierarchical system. A supply-oriented dynamic approach, *Regional Science and Urban Economics*, 16(1), pp. 145–160.

Champion, A. G. (2001) A changing demographic regime and evolving polycentric urban regions-consequences for the size, composition and distribution of city populations, *Urban Studies*, 38(4), pp. 657–677.

Chen, Y. & Rosenthal, S. S. (2008) Local amenities and life-cycle migration: Do people move for jobs or fun? *Journal of Urban Economics*, 64(3), pp. 519–537.

Christaller, W. (1933) *Die Zentralen Orte in Süddeutschland* (Jena: Gustav Fischer).

Clark, T. N. (2004) *The City as an Entertainment Machine* (Amsterdam: Elsevier–JAI Press).

Clark, T. N., Lloyd, R., Wong, K. K. & Jain, P. (2002) Amenities drive urban growth, *Journal of Urban Affairs*, 24(5), pp. 493–515.

Cook, D. O., Kieschnick, R. & McCullough, B. D. (2008) Regression analysis of proportions in finance with self-selection, *Journal of Empirical Finance*, 15(5), pp. 860–867.

Dahlberg, M., Eklöf, M., Fredriksson, P. & Jofre-Monseny, J. (2012) Estimating preferences for local public services using migration data, *Urban Studies*, 49(2), pp. 319–336.

Dalmazzo, A. & De Blasio, G. (2011) Amenities and skill-biased agglomeration effects: Some results on Italian cities, *Papers in Regional Science*, 90(3), pp. 503–527.

De Goei, B., Burger, M. J., Van Oort, F. G. & Kitson, M. (2010) Functional polycentrism and urban network development in the Greater South East, United Kingdom: Evidence from Commuting Networks, 1981–2001, *Regional Studies*, 44(9), pp. 1149–1170.

Eaton, B. C. & Lipsey, R. G. (1982) An economic theory of central places, *Economic Journal, Royal Economic Society*, 92(365), pp. 56–72.

Dobkins, L. H. & Ioannides, Y. M. (2001) Spatial interactions among U.S. cities: 1900–1990, *Regional Science and Urban Economics*, 31(6), pp. 701–731.

Ebertz, A. (2013) The capitalization of public services and amenities into land prices—empirical evidence from German communities. *International Journal of Urban and Regional Research*, 37(6), pp. 2116–2128.

Ferrari, S. L. P. & Cribari-Neto, F. (2004) Beta regression for modelling rates and proportions, *Journal of Applied Statistics*, 31(7), pp. 799–815.

Florida, R., Gulden, T. & Mellander, C. (2008) The rise of the mega-region, *Cambridge Journal of Regions, Economy and Society*, 1(3), pp. 459–476.

Fujita, M., Krugman, P. & Venables, A. J. (1999) *The Spaital Economics: Cities, Regions and International Trade* (Cambridge, MA: MIT Press).

Giffinger, R. & Suitner, J. (2014) Polycentric metropolitan development: From structural assessment to processual dimension, *European Planning Studies*, doi:10.1080/09654313.2014.905007

Glaeser, E. L., Kolko, J. & Saiz, A. (2001) Consumer city, *Journal of Economic Geography*, 1(1), pp. 27–50.

Hall, P. & Pain, K. (Eds) (2006) *The Polycentric Metropolis: Learning from Mega-City Regions in Europe* (London: Earthscan).

Hoyler, M., Kloosterman, R. C. & Sokol, M. (2008) Polycentric puzzles—emerging mega-city regions seen through the lens of advanced producer services, *Regional Studies*, 42(8), pp. 1055–1064.

Kieschnick, R. & McCullough, B. D. (2003) Regression analysis of variates observed on (0,1): Percentages, proportions and fractions, *Statistical Modelling*, 3(3), pp. 193–203.

Kloosterman, R. C. & Musterd, S. (2001) The polycentric urban region: Towards a research agenda, *Urban Studies*, 38(4), pp. 623–633.

Lee, S. (2010) Ability sorting and the consumer city, *Journal of Urban Economics*, 68(1), pp. 20–33.

Markusen, A. & Schrock, G. (2009) Consumption-driven urban development, *Urban Geography*, 30(4), pp. 1–24.

Meijers, E. (2007a) Clones or complements? The division of labour between the main cities of the Randstad, the Flemish Diamond and the Rheinruhr Area, *Regional Studies*, 41(7), pp. 889–900.

Meijers, E. (2007b) From a central place to a network model: Theory and evidence of a paradigm change, *Tijdschrift voor Economische en Sociale Geografie*, 98(2), pp. 245–259.

Meijers, E. (2008) Summing small cities does not make a large city: Polycentric urban regions and the provision of cultural, leisure and sports amenities, *Urban Studies*, 45(11), pp. 2323–2342.

Mulligan, G. F. (1984) Agglomeration and central place theory: A review of the literature, *International Regional Science Review*, 9(1), pp. 1–42.

Mushinski, D. & Weiler, S. (2002) A note on the geographical interdependencies of retail market areas, *Journal of Regional Science*, 42(1), pp. 75–86.

Ospina, R. & Ferrari, S. L. P. (2012) A general calss of zero-or-one inflated beta regression models, *Computational Statistics and Data-Analysis*, 56(6), pp. 1609–1623.

Papke, L. E. & Woolridge, J. M. (1996) Econometric methods for fractional response variables with an application to 401(K) plan participation rates, *Journal of Applied Econometrics*, 11(6), pp. 619–632.

Papke, L. E. & Woolridge, J. M. (2008) Panel data methods for fractional response variables with an application to test pass rates, *Journal of Econometrics*, 145(1–2), pp. 121–133.

Parr, J. B. (2004) The polycentric urban region: Acloser inspection, *Regional Studies*, 38(3), pp. 231–240.

Partridge, M. D., Rickman, D. S., Ali, K. & Olfert, M. R. (2009) Do new economic geography agglomeration shadows underlie current population dynamics across the urban hierarchy? *Papers in Regional Science*, 88(2), pp. 445–466.

Peck, J. (2005) Struggling with the creative class, *International Journal of Urban and Regional Research*, 29(4), pp. 740–770.

Peeters, D. (2011) *The Functional Urban Areas Database* (Luxembourg: ESPON). Available at http://www.espon.eu (accessed 29 April 2013).

Phelps, N. A. & Ozawa, T. (2003) Contrasts in agglomeration: Proto-industrial, industrial and post-industrial forms compared, *Progress in Human Geography*, 27, pp. 583–604.

Rappaport, J. (2008) Consumption amenities and city population density, *Regional Science and Urban Economics*, 38(6), pp. 533–552.

Roback, J. (1982) Wages, rents, and the quality of life, *Journal of Political Economy*, 90, pp. 1257–1278.

Rosen, S. (1979) Wage-based indexes of urban quality of life, in: P. N. Miezkowski and M. R. Straszheim (Eds) *Current Issues in Urban Economics*, pp. 74–104 (Baltimore, MD: Johns Hopkins University Press).

Rosenthal, S. S. & Strange, W. C. (2004) Evidence on the nature and sources of agglomeration economies, in: J. V. Henderson & J. F. Thisse (Eds) *Handbook of Regional and Urban Economics*, 1st ed., Vol. 4, Chapter 49, pp. 2119–2171 (Amsterdam: Elsevier).

Schmid, M., Wickler, F., Maloney, K. O., Mitchell, R., Fenske, N. & Mayr, A. (2013) Boosted beta regression, *PLOS ONE*, 8(4), pp. 1–15.

Silver, D. (2012) The American scenescape: Amenities, scenes and the qualities of local life, *Cambridge Journal of Regions, Economy and Society*, 5(1), pp. 97–114.

Smidt-Jensen, S., Skytt, C. B. & Winther, L. (2009) The geography of the experience economy in Denmark: Employment change and location dynamics in attendance-based industries, *European Planning Studies*, 17(6), pp. 847–862.

Spiekermann, K. & Wegener, M. (2006) Accessibility and spatial development in Europe, *Scienze Regionali*, 5(2), pp. 15–46.

Storper, M. & Scott, A. J. (2009) Rethinking human capital, creativity and urban growth, *Journal of Economic Geography*, 9(2), pp. 147–167.

Tabuchi, T. & Yoshida, A. (2000) Separating agglomeration economies in consumption and production, *Journal of Urban Economics*, 48(1), pp. 70–84.

67

Thilmany, D., McKenney, N., Mushinski, D. & Weiler, S. (2005) Beggar-thy-neighbor economic development: A note on the effect of geographic interdependencies in rural retail markets, *The Annals of Regional Science*, 39(3), pp. 593–605.

Van der Laan, L. (1998) Changing urban systems: An empirical analysis at two spatial levels, *Regional Studies*, 32(3), pp. 235–247.

Van Duijn, M. & Rouwendal, J. (2013) Cultural heritage and the location choice of Dutch households in a residential sorting model, *Journal of Economic Geography*, 13(3), pp. 473–500.

Van Oort, F. G., Burger, M. J. & Raspe, O. (2010) On the economic foundation of the urban network paradigm. Spatial integration, functional integration and urban complementarities within the Dutch Randstad, *Urban Studies*, 47(4), pp. 725–748.

Wensley, M. R. D. & Stabler, J. C. (1998) Demand-threshold estimation for business activities in rural Saskatchewan, *Journal of Regional Science*, 38(1), pp. 155–177.

Appendix 1. Zero-and-one inflated beta regression on Cultural Amenities Index—interaction effects

	Coefficients		
	Model 4 (Graph 4)	Model 5 (Graph 5)	Model 6 (Graph 6)
Proportion part			
Local population (100 k)	0.389 (.103)**	0.202 (.044)**	0.247 (.034)**
Population rest FUA (mln)	−0.020 (.021)	−0.031 (.025)	−0.015 (.021)
(Inter)national accessibility (10 mln)	0.073 (.031)*	0.059 (.030)*	−0.015 (.026)
GDP per capita FUA (k)	0.020 (.008)**	0.019 (.008)*	0.023 (.008)**
Dummy heritage site	0.594 (.190)**	0.588 (.186)**	0.630 (.190)**
Dummy not largest location in FUA	0.787 (.093)**	0.692 (.087)**	−0.441 (.317)
Dummy not largest location in FUA*Local population (100 k)	−0.131 (.110)		
Dummy not largest location in FUA*Population rest FUA (mln)		1.298 (.251)**	0.176 (.046)**
Dummy not largest location in FUA*(Inter)national accessibility (10 mln)			
Zero-inflated part			
Local population (100 k)	−4.534 (.408)**	−4.491 (.364)**	−4.562 (.342)**
Population rest FUA (mln)	−0.038 (.045)	−0.039 (.044)	−0.044 (.044)
(Inter)national accessibility (10 mln)	0.214 (.061)**	0.217 (.061)**	0.259 (.063)**
GDP per capita FUA (k)	0.008 (.012)	0.008 (.012)	0.008 (.012)
Dummy heritage site	−0.659 (.397)	−0.665 (.398)	−0.680 (.408)
Dummy not largest location in FUA	−0.299 (.330)	−0.109 (.226)	1.427 (.712)*
Dummy not largest location in FUA*Local population (100 k)	0.187 (.681)		
Dummy not largest location in FUA*Population rest FUA (mln)		−7.251 (6.56)	−0.279 (.118)*
Dummy not largest location in FUA*(Inter)national accessibility (10 mln)			
Number of observations	2794	2794	2794
Wald chi-square	538.4**	665.4**	600.2**
ln phi	4.06 (.135)**	4.19 (.122)**	4.11 (.135)**
Country dummies +	Yes	Yes	Yes

Notes: Robust standard errors in parentheses. All models are estimated with intercept. The one-inflated part is not displayed as it is an intercept-only model. Since average marginal effects for the interaction effects are not provided, these are not displayed.
*$p < .05$.
**$p < .01$.

Related Variety and Regional Economic Growth in a Cross-Section of European Urban Regions

FRANK VAN OORT*, STEFAN DE GEUS** & TEODORA DOGARU[†,‡]

*Urban and regional Research Centre Utrecht, Utrecht University, Utrecht, The Netherlands, **Triodos Investment Management, Triodos Bank, Zeist, The Netherlands, [†]Department of Economic Analysis and Business Administration, University of A Coruna, A Coruna, Spain, [‡]Faculty of Geosciences, Utrecht University, Utrecht, The Netherlands

ABSTRACT *This paper introduces indicators of regional related variety and unrelated variety to conceptually overcome the current impasse in the specialization-diversity debate in agglomeration economics. Although various country-level studies have been published on this conceptualization in recent years, a pan-European test has been missing from the literature until now. A pan-European test is more interesting than country-level tests, as newly defined cohesion policies, smart-specialization policies, place-based development strategies and competitiveness policies may be especially served by related variety and unrelated variety conceptualizations. We test empirically for the significance of variables based on these concepts, using a cross-sectional data set for 205 European regions during the period 2000–2010. The results confirming our hypotheses are that related variety is significantly related to employment growth, especially in small and medium-sized city-regions, and that specialization is significantly related to productivity growth. We do not find robust relationships that are hypothesized between unrelated variety and unemployment growth.*

1. Introduction

This paper focuses on agglomeration circumstances influencing economic growth across European urban regions. Empirical studies on agglomeration economies are characterized by a high diversity of approaches. Rosenthal and Strange (2004) present a brief review of papers focusing on urbanization economies as advantages of cities applying to every firm or consumer. Noteworthy is that most early (pre-1990s) works on agglomeration simply used cities' population as a measure of agglomeration. These studies assume that the population elasticity of productivity is constant. Rosenthal and Strange (2004) conclude that this literature has found relatively consistent evidence: doubling the population of a city increases productivity by 3–8%. After the findings of Glaeser *et al.* (1992), who studied sectoral agglomeration effects more than the aggregated effect, it has become more commonplace to analyse growth variables using employment in cities, suggesting a relationship between agglomeration and economic growth and thereby introducing the possibility that increasing returns in an urban context operate in a dynamic, rather than static, context (Beaudry & Schiffauerova, 2009; De Groot *et al.*, 2009; Melo *et al.*, 2009). Sector-specific localization economies, stemming from input–output relations and firms' transport cost savings, human capital externalities and knowledge spill-overs, are generally offset against the general urbanization economies. A large body of literature builds on this new conceptualization of agglomeration economies, as reflected in three recent overviews and meta-studies. These studies show that the relation between agglomeration and growth is ambiguous and indecisive with regard to whether specialization or diversity is facilitated by (sheer) urbanization as context. The first goal of this paper is to take a step towards the concept of renewal as a possible way out of this currently seemingly locked-in debate and to introduce related variety and unrelated variety as concepts in the empirical modelling of growth across European regions. These concepts have been tested only at the country level in Europe,[1] and no pan-European test has been provided due to data limitations. This paper provides a first pan-European test of these concepts.

The second goal of the paper is to contribute to the recent policy discussion on place-based or place-neutral development strategies in the European Union (EU). This debate is highlighted in the context of a series of recent major policy reports: the place-neutral policies in the 2009 World Bank report (World Bank, 2009) and the European place-based development strategies in the studies of Barca (2009) and Barca *et al.* (2012). As highlighted by Van Oort and Bosma (2013) and Barca *et al.* (2012), place-neutral strategies rely on the agglomerative forces of the largest cities and metropolitan regions to attract talent and growth potential. Place-based development strategists claim that the polycentric nature of a set of smaller and medium-sized cities in Europe (often also called second-tier cities), each with its own peculiar characteristics and specialization in the activities to which it is best suited, creates fruitful urban variety, which optimizes economic development. This perspective implies that medium-sized city-regions have not declined in importance relative to larger urban ones, a proposition that has indeed been indicated in monitoring publications by OECD (2009, 2011, 2012). Until now, however, there has been little empirical support for explanations based on the concepts of related variety and unrelated variety and sectoral specialization.

For our empirical testing, we use a fairly standardized setup entailing the cross-sectional modelling of agglomeration externalities and economic growth (employment growth and productivity growth between 2000 and 2010) that distinguishes among various drivers of localized growth processes. In line with Dogaru *et al.* (2011), we show that this type of modelling is informative for competitive and cohesive growth policies in the EU, especially those with a focus on the role of medium-sized urban regions.

This paper is structured as follows. Section 2 introduces the current locked-in debate on specialization versus diversity dominance in agglomeration economics. This section shows that the dominance of neither concept can be established and that conceptual renewal is needed to move past this controversy. Section 3 introduces related variety and unrelated variety as such a conceptual renewal to act in empirical modelling instead of generalized indicators for urbanization economies that have usually been applied in modelling. This section ends with three hypotheses concerning the effects of related variety, unrelated variety and sectoral specialization on regional economic development. A fourth hypothesis is formulated based on smaller and medium-sized European regions in relation to economic development. Section 4 introduces the data and variables used in our regional cross-sectional growth models. Section 5 presents our modelling outcomes for regional employment growth, productivity growth and unemployment growth for the period 2000–2010. Section 6 concludes and directly addresses the four hypotheses and the issues of place-based development strategies.

2. Agglomeration Economies Between Specialization and Diversity

Agglomeration economies in relation to urban and regional growth are receiving attention in an ever-burgeoning literature on its causes, magnitude and (policy) consequences. This rise of agglomeration economies in economic and geographical studies has met much criticism (McCann & van Oort, 2009). Some observers have argued that the modern treatment of agglomeration economies and regional growth in fact represents a rediscovery by economists of well-rehearsed concepts and ideas with a long history in economic geography. Several criticisms of the monopolistic modelling logic underpinning New Economic Geography have come from economic geography schools of thought and from both (neo)classical and heterodox schools of economics. Conversely, advocates of relatively new economic approaches, such as institutional economics and evolutionary economic geography, argue that their analyses provide insights into spatial economic phenomena that were previously unattainable under existing analytical frameworks and toolkits.

A prime example of potential gains of different theories and conceptual frameworks is the specialization-diversity debate in the urban economics and economic geography literatures. Should regions and cities specialize in certain products or technologies to locally gain from economies of scale (in so-called clusters), shared labour markets and input–output relations, or should regions diversify over various products and industries and hence have both growth opportunities from inter-industry spill-overs as well as portfolio advantages that hedge a regional economy in times of economic turmoil? This question has captured the attention of many researchers over the last two decades, following papers by Glaeser *et al.* (1992) and Henderson *et al.* (1995) that advocate sectoral diversity and specialization, respectively, as the main economic–geographic circumstance propagating growth. The dichotomy of specialization-diversity has ever since been treated as a rather strict division—many studies try to find the definitive answer to the question "Who is right: Marshall or Jacobs?" (quoted from the title of Beaudry & Schiffauerova, 2009). Although practically every study conducted in the framework tries to conclude that either specialization or diversity is a driver of growth and innovation, the studies by Van Oort (2004), Paci and Usai (1999), Neffke *et al.* (2011) Shefer and Frenkel (1998), Duranton and Puga (2000) and O'Hualloachain and Lee (2011) prove that this is in fact not an "either–or" question, finding that both specialization and diversity matter for regional economic performance—on different geographical levels, for different time periods, over the industry lifecycle and in different institutional settings.

That the specialization-diversity issue is not an "either–or" question has now been concluded by two meta-studies and an extensive overview of all published empirical analyses on this matter (Beaudry & Schiffauerova, 2009; De Groot *et al.*, 2009; Melo *et al.*, 2009).

From these three overviews, it becomes clear that the specialization-diversity debate appears to become an unproductive line of argument in addressing the nature, magnitude and determinants of agglomeration externalities (see also Desrochers & Leppälä, 2011). The answer to the "either–or" diversity-specialization question is at best inconclusive, with outcomes being dependent on measurement in many respects (e.g. scale, composition, context, period, type of performance indicators). Aside from these methodological issues, the many tests provided do not actually measure knowledge transfer or knowledge spill-overs (Van Oort & Lambooy, 2014)—one of the main mechanisms supposed to drive agglomeration economies. Finally, theoretically the debate focuses on the old theory of agglomeration as introduced by Marshall (1890) and does not use insights from newly developed theoretical models and conceptualizations in evolutionary and institutional geographical approaches.

3. Conceptual Renewal and Hypotheses: Related Variety and Unrelated Variety, Specialization and Place-Based Development

The divergence observed in the literature concerning diversification and specialization, in addition to the observed differences in measurements of classifications and methodological issues, is most likely related to the weak conceptualization and limited theoretical underpinning of the concepts. New theoretical developments in institutional and evolutionary economic geography have recently emerged, offering heterodox economic explanations for the regional economic development and the role of relatedness and diversification (Boschma & Martin, 2010). For economic geographers, as well as institutional and evolutionary economists working in this tradition, cultural and cognitive proximity are deemed to be equally as important as geographical proximity in the transmission of ideas and knowledge (Boschma, 2005). Boschma and Lambooy (1999) further argue that the generation of local externalities is also crucially linked to the importance of variety and selection in terms of the "fitness" of a local milieu. The now-burgeoning tradition in evolutionary economic geography has prompted the question of whether concepts of diversification and specialization may fully capture the complex role of variety within the capitalist economy. This development has led to a recent revival of interest in the role of specific forms of variety, specifically related variety and unrelated variety (Frenken *et al.*, 2007). Jacobs (1969) initiated the idea that the variety of a region's industry or technological base may affect economic growth. Frenken *et al.* (2007) state that variety and diversification consist of related variety and unrelated variety, arguing that not simply the presence of different technological or industrial sectors will trigger positive results but that sectors require complementarities that exist in terms of shared competences. This need induces a distinction in related variety and unrelated variety because knowledge spill-overs will not transfer to all different industries evenly, due to the varying cognitive distances between each pair of industries. It is argued that industries are more highly related when they are closer to each other within the Standard Industrial Classification (SIC) system. Frenken *et al.* (2007) find that for Dutch urban regions, the positive results of knowledge spill-overs are higher in regions with related variety, whereas regions characterized by unrelated variety are better hedged for economic shocks (portfolio effect). The authors also find marked differences between employment growth and productivity growth. An interesting theoretical contribution to the specialization-variety debate that focuses on these explained variables has been provided by lifecycle theory, which holds that industry evolution is characterized by product innovation (and more employment growth) in a first stage and process innovation (and more productivity growth) in a second stage. This distinction does not imply that product innovation occurs exclusively at the time of birth of a new industry, with process innovation only occurring thereafter. Rather, product lifecycle theory assumes that product innovation peaks before process innovation peaks. In accordance with the economics of agglomeration, evolutionary economists also stress the important role of variety in creating new varieties. In other words, Jacobs' externalities are assumed to play an important role in urban areas in creating new varieties, new sectors and employment growth. When firms survive and become mature, they tend to standardize production and become more capital-intensive and productive.

This background leads to three hypotheses on the relation between specialization, variety and economic development in regions:

Hypothesis 1: Urban regions with a sector structure of *related variety* experience an increased rate of product innovation, co-evolving with higher *employment* in the short run. We summarise this hypothesis as follows to create a testable version for this paper: *In the short run, employment growth is positively related to related variety and negatively related to specialisation.*

Hypothesis 2: Regions with a sector structure of *unrelated variety* experience fewer job losses from asymmetric shocks, which leads to *lower unemployment growth.* We summarise this hypothesis as follows to create a testable version for this paper: In the short run, unemployment growth is negatively related to unrelated variety.

Hypothesis 3: Regions with a specialised sector structure experience an increased rate of process innovation and reduced production costs, which potentially leads to higher *productivity growth.* This phenomenon is more pronounced in the short run than in the long run. We summarise this hypothesis as follows to create a testable version for this paper: *In the short run, labour productivity growth is positively related to specialisation.*

A fourth hypothesis relates the agglomeration concepts to urban population size of regions, and may be indirectly linked to urban structure. In Europe, the character of urban regions is fundamentally different from that of urban regions in other parts of the world (such as the US and Asia). It is exactly this urban structure that has fuelled the recent place-based versus place-neutral development debate. Barca *et al.* (2012) and Van Oort and Bosma (2013) summarize the place- and people-based policy debate in the European context in detail. Based on current economic geographical theories of innovation and the density of skills and human capital in cities, globalization and endogenous growth through urban learning opportunities, spatially blind approaches argue that intervention, regardless of context, is the best way to resolve the old dilemma of whether development should be about "places" or "people" (Barca *et al.*, 2012, p. 140). It is argued that agglomeration in combination with encouraging people's mobility not only allows individuals to live where they expect to be better off but also increases individual incomes, productivity, knowledge and aggregate growth (World Bank, 2009). Consequently, development intervention should be space-neutral, and factors should be encouraged to move to where they are most productive. In reality, this phenomenon occurs primarily in large cities and city-regions. In contrast, the place-based approach assumes that the interactions between institutions and geography are critical for development, and many of the clues for development policy lie in these interactions. Investigating the interactions between institutions and geography to understand the likely impacts of a policy requires the explicit consideration of the specifics of the local and regional contexts (Barca et al., 2012, p. 140). The various forms that proximity may take in networks (e.g. physical, social, technological and institutional) are important in this respect (Thissen et al., 2013).

According to place-based development strategists, economic growth is not uniquely related to mega-city-regions (Barca et al., 2012). Instead, growth may be distributed across various urban systems in different ways in different countries (OECD, 2009, 2011). The place-based approach's emphasis on interactions between institutions and economic geography has allowed for the examination of development in European regions of all sizes (Dijkstra et al., 2013). Because the roles of very large and small communities have been addressed extensively in the literature (Dijkstra et al., 2013), Barca et al. (2012) emphasize the simultaneous role of medium-sized ("second-tier") urban regions and argue that these are over-represented in Europe. Many highly productive urban regions in the EU are indeed small- to medium-sized, and their dominant competitive advantage is that they exhibit high degrees of connectivity compared to urban or home market scales (ESPON, 2012). This phenomenon leads to the formulation of our fourth hypothesis:

Hypothesis 4: Agglomeration externalities are related to economic performance in all sizes of urban regions in Europe.

4. Data and Variables Used in Empirical Analysis

Our empirical analysis tests the relationship between productivity growth and employment growth in distinctive large, capital regions in Europe on the one hand and medium-sized and small urban regions on the other, controlling for other important factors, and makes conclusions on the place-based policy implications suggested in the recent policy discourse. To test our hypotheses, we conduct an empirical analysis on growth differentials over 205 European NUTS2 regions in 15 EU countries[2] between 2000 and 2010, focusing on different population sizes of regions.

Measuring diversification over sectors in regional economies is sensitive to the indicator applied. In our empirical analysis, we apply an entropy measure (see Frenken et al., 2007 for a detailed discussion). The main advantage of the entropy measure, and the reason for its use in the context of diversification, is that entropy can be decomposed at each sectoral digit level. The decomposable nature of entropy implies that variety at several digit levels can enter a regression analysis without necessarily causing collinearity. In the context of measuring regional variety to analyse the effects on growth, decomposition is informative, as one expects entropy/variety at a high level of sector aggregation to have a portfolio effect on the regional economy, protecting it from unemployment, whereas one expects entropy/variety at a low level of sector aggregation to generate knowledge spill-overs and employment growth. Put differently, entropy at a high level of sector aggregation measures unrelated variety, whereas entropy at a low level of sector aggregation measures related variety. We use geo-coded AMADEUS micro data (provided by Bureau van Dijk) on European firms aggregated into European NUTS2 regions as a source for the calculation of related variety and unrelated variety. Because small firms are underrepresented in this database, firm-level data are weighted by turnover values (inverse). This approach allows us to best capture the large and sectorally heterogeneous regional economy. Marginal variety may be computed at all four-digit SIC levels in the data set, indicating an increase in variety when moving from one-digit level to the next. Because the marginal entropy levels at the three- and four-digit levels are correlated strongly, we chose to compute the marginal increase when moving from the one-digit level to the four-digit level. We label this variety indicator as *related variety*, as opposed to the two-digit level entropy, which we associate with unrelated variety. We will include both types of variety to test whether related variety and unrelated variety have different effects.

Formally, all four-digit sectors i fall exclusively under a two-digit sector S_g, where $g = 1, \ldots, G$, one can derive the two-digit shares P_g by summing the four-digit shares p_i

$$P_g = \sum_{i \in S_g} p_i. \tag{1}$$

The entropy at the two-digit level, or unrelated variety, is given by

$$UV = \sum_{g=1}^{G} P_g \log_2 \left(\frac{1}{P_g} \right). \tag{2}$$

And, the weighted sum of entropy within each one-digit sector is given by

$$RV = \sum_{g=1}^{G} P_g H_g, \tag{3}$$

where

$$H_g = \sum_{i \in S_g} \frac{p_i}{P_g} \log_2 \left(\frac{1}{p_i / P_g} \right), \quad g = 1, \ldots, G. \tag{4}$$

As argued earlier, we consider related variety to be the indicator for Jacobs externalities because it measures the variety *within* each of the two-digit classes. We expect the economies arising from variety to be especially strong between sub-sectors, as knowledge spills over primarily between firms selling related products. By contrast, unrelated variety measures the extent to which a region is diversified in very different types of activity. This type of variety is expected to be instrumental in avoiding unemployment.

The maps of related variety and unrelated variety in European regions are provided in Figures 1 and 2, which present two different regional patterns for unrelated variety (between two-digit variety) and related variety (marginal increase in entropy when moving from one- to four-digit differences, so within one-digit variety). As the maps clearly show, variety at high levels of aggregation shows no strong resemblance with variety at low levels, which strongly suggests that the choice of sector aggregation is not trivial. Related variety appears to be a somewhat more *urban* regional feature than unrelated variety (compare Frenken *et al.*, 2007).

Employment and labour productivity (output per employee) data were obtained from the Cambridge Econometrics Statistical Database on European regions. We obtained

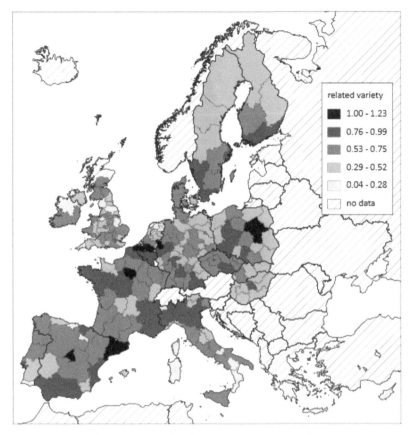

Figure 1. Related variety across European regions.

data from this data set for the years 2000 and 2010. Productivity growth and employment growth are defined as $\ln(\text{emp}_{2010}/\text{emp}_{2000})$ and $\ln(\text{prod}_{2010}/\text{prod}_{2000})$ to normalize their distributions.

Localization economies, measured in 2000 for endogeneity reasons, are associated with the concentration of a particular sector in a region. This type of economy is often captured using specialization indicators. The degree of regional specialization in our models is measured using the Theil Index over the location quotients of production in 59 products, including agriculture, manufacturing and services. This unique data set has been collected by the Netherlands Environmental Assessment Agency (for a description see Thissen *et al.*, 2013) and is based on regionalized production and trade data for European nuts2 regions, 14 sectors and 59 product categories (compare Combes & Overman, 2004). Location quotients measure the relative concentration sectors in a region as the percentage of employment accounted for by a sector in a region relative to the percentage of employment accounted for by that sector in Europe as a whole. This quotient measures whether a sector is over- or underrepresented in a region compared with its average representation in a larger area and thus is to comprise localization or specialization economies of agglomeration. The Theil coefficient then measures deviations from the European average distribution of employment specializations in all

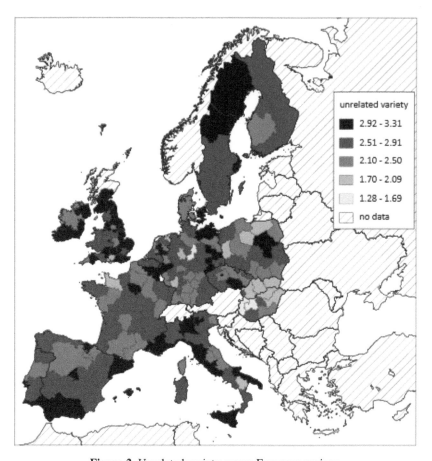

Figure 2. Unrelated variety across European regions.

sectors. This transformation transforms the individual sectoral concentration measures in a generalized specialization measure. A high score represents a large degree of sectoral specialization in a region, and a low score represents sectoral diversity. In the largest national economies of Germany, France and the UK, regions have high levels of sectoral diversity (all regions contain most of the existing sectors, including services). Eastern European regions in Poland, Slovakia, the Czech Republic and Hungary are relatively specialized, as are Scandinavian and Irish regions. These regions lack concentrations of certain activities, e.g. specific types of services, manufacturing, distribution or agricultural activities. A group of medium-sized economies, such as the Netherlands, Belgium, Denmark, Italy, Portugal and Spain, show moderate levels of specialization.

To test and control for either convergence or divergence, both productivity and employment growth in the period 2000–2010 are, respectively, related to the productivity and employment levels in 2000. These relations are hypothesized to be negative (convergence). All other explanatory variables in our models for employment and productivity growth are also measured for the year 2000 because of endogeneity reasons. The circumstances shown in 2000 may cause subsequent growth in the period 2000–2010, but those shown in 2010 cannot. *Investments in private and public research and development (R&D)*

are calculated as percentages of gross domestic product (GDP) from Eurostat statistics. These investments in innovation are generally believed to be positively related to economic growth (Moreno *et al.*, 2005). Private R&D investments occur mainly in regions with larger multinational enterprises. Public R&D is more highly related to regions with technological universities and regions where universities and firms ally.

The *degree of economic openness* of European regions is calculated as the total value of imports and exports in a region divided by the region's GDP. This indicator for the volume of trade is based on a make-and-use table (input-output-table) for 2000 at the nuts2 level concerning 14 sectors and 59 product categories, including services. This data set is developed by the Netherlands Environmental Assessment Agency. The volume of trade increases with the size of the region at a declining rate and is strongly dependent on global economic development with competition in global markets, driving up productivity and attracting new investments and collaborations. High potential may also spill over to nearby regions or in the regional network of specialized and subcontracting industries and regions. *Density* (measured as population density) measures whether agglomeration (economic size) plays a role in economic growth. This dimension of agglomeration is related not to localization economies (specialization) and diversity economies but to pure urban size effects (Frenken *et al.*, 2007). In general, the literature suggests that higher density enables better interaction, enhancing growth (Puga, 2002).

We measured the *average educational level* of regions by the percentage of the tertiary and higher educated population within the total population. The relationship of education with (employment and productivity) growth is thought to be positive, as more highly skilled people can be more productive, and agglomeration may attract more of these people. Remarkably low scores on this indicator are found in eastern European regions and Italian regions. The regional *wage level*, as an indicator of personal income, is hypothesized to be positively related to growth. The wage level variable is highly correlated with GDP per capita as an indicator. Higher wage levels and productivity levels are also highly correlated. In the productivity growth models, the wage level is thus excluded from the analysis (and the productivity level is included). *Market potential* is measured by a gravity equation on production in all regions, corrected for distances. Finally, a dummy variable is introduced into the models for large and capital regions opposed to medium-sized and smaller regions. The degree of urbanization over the 205 regions is determined by the *distribution of classes* distinguished in OECD (2012, 2013) comprising large and capital regions (at least 3 million inhabitants), medium-sized regions (between 1.5 and 3 million inhabitants) and small regions (fewer than 1.5 million inhabitants). Although this distinction differs from the one originally presented for all cities in the world in OECD (2012, 2013), these cut-off points yield a distribution for the European regional classification adopted in this paper that is comparable to the OECD distribution on a global scale. In our analysis, large and capital regions are categorized within the large urban regime, and small- and medium-sized regions are categorized within the medium-sized urban regime.

To avoid multicollinearity in our models, we tested for high correlations among these explanatory variables, and we analysed variance inflation factors for each variable added to the models. None of the correlations is disturbingly high. As previous research has shown that spatial dependence between proximate regions in Europe is an important source for divergent growth opportunities in productivity and employment (Le Gallo *et al.*, 2011), we control for this finding in our analyses by introducing maximum likelihood (ML) estimation, which includes spatial lags, using inverse distance weighting matrices (noted W_1) and squared inverse distance matrices (noted W_2). Inverse distances are calculated from core to core without cut-off point.

5. Modelling Outcomes

This section discusses modelling outcomes presented in Tables 1–3 for employment growth, productivity growth and unemployment growth, respectively. The models are constructed in similar ways, starting with a ML spatial-lag model that corrects for spatial dependence (the spatial-lag variable is denoted as w_growth in the tables), moving into a ML-spatial-lag model that decomposes the observation into the following two regimes, which are estimated simultaneously and which we wish to use to test our hypotheses: the regime of large and capital regions and the regime of medium-sized and smaller regions. The model fit usually increases over these successive modelling steps, the significance of spatial regimes are indicated by the outcomes of a spatial Chow–Wald test, and variables that significantly differ from each other over regimes are presented in boxes in the tables. In all three models the Breusch-Pagan (BP) tests indicate problems of heteroskedasticity, which makes our interpretation cautious. Due to space limitations, we focus on our four hypotheses in our discussion of the outcomes. Hypothesis 1 links related variety to employment growth positively and to specialization negatively. Table 1 shows that this hypothesis is confirmed in the spatial-lag models (2) and (3). The regime analysis in column (3) shows that this relationship particularly holds true for medium-sized and smaller regions and not for larger capital regions. Hypothesis 2 positively linked specialization to productivity growth. Table 2 confirms this for all models and all regimes applied, indicating that this relation is very robust.

The regime analysis shows that large urban and capital regions feature a stronger relationship between specialization and productivity growth compared to medium-sized regions. The third hypothesis proposes that unemployment growth is negatively related to unrelated variety (portfolio argument). The findings presented in Table 3 indicate that this hypothesis is rejected for all specifications. The reasons for this finding may be diverging national regulations and institutions in Europe, which cause national regimes to exist across the continent. This finding also indicates that for this variable, pan-European relations highly diverge from those found at individual country levels. Our fourth hypothesis, stating that regions of all sizes are involved in growth accounting, is confirmed. Employment growth is more naturally suited in medium-sized regions, whereas productivity growth is enabled by specialization patterns in both large and medium-sized regions (with a higher coefficient being found in large urban regions). When Jacobs' externalities are an important ingredient for employment growth in small and medium-sized urban regions, this means that these regions are well equipped for creating new varieties and attract new and related sectors. This economical vital role of medium-sized and small urban regions has not been suggested previously. Perhaps due to agglomeration disadvantages, the largest urban regions do not automatically show this dynamics.

6. Conclusions and Discussion

This paper introduces indicators of regional related variety and unrelated variety to conceptually overcome the current impasse in the specialization-diversity debate in agglomeration economics. Although various country-level studies have been introduced on this

Table 1. Modelling outcomes for employment growth 2000–2010

Explanatory variables	(1) Spatial-lag model W_1		(2) Spatial-lag model W_2		(3) Regimes' urban size Small- and medium-sized		Large and capital regions	
(Constant)	0.30	0.19	0.31*	0.16	0.71**	0.21	−0.15	0.54
Employment 2000	−0.03**	0.01	−0.03**	0.01	−0.04**	0.01	0.03	0.05
Private R&D	0.00	0.01	0.00	0.01	0.00	0.01	−0.01	0.03
Public R&D	0.00	0.01	0.00	0.01	0.00	0.01	−0.02	0.02
Openness economy	0.05**	0.01	0.03**	0.01	0.02	0.01	0.06	0.06
Market potential	−0.07**	0.02	−0.05**	0.02	−0.05**	0.02	−0.04	0.06
Education	0.02	0.01	0.01	0.01	0.01	0.01	0.04	0.07
Population density	0.01	0.01	0.01*	0.01	0.01	0.01	−0.02	0.02
Wages	0.03**	0.01	0.02**	0.01	0.01	0.01	0.01	0.03
Related variety	0.09**	0.03	0.09**	0.03	0.11**	0.03	−0.07	0.12
Unrelated variety	0.03**	0.01	0.02**	0.01	0.02	0.01	0.07	0.08
Specialization	−0.28**	0.11	−0.19**	0.09	−0.24**	0.09	−0.22	0.37
W_Employment growth	0.95**	0.04	0.92**	0.04	0.92**		0.039	
Summary statistics								
N	205		205		205			
R²	0.291		0.402		0.447			
Chow–Wald	—		—		18.6		0.1	
BP (heteroskedasticity)	42.002	0	45.915	0	3.565		0.059	
LR (spatial-lag)	37.583	0	80.966	0	83.4		0	
LM (spatial error)	48.364	0	2484	0.115	1.042		0.307	

Notes: Coefficients and *t*-values.

*p < .10.

**p < .05.

Table 2. Modelling outcomes for productivity growth 2000–2010

Explanatory variables	(1) Spatial-lag model W_1		(2) Spatial-lag model W_2		(3) Regimes' urban size			
					Small- and medium-sized		Large and capital regions	
(Constant)	0.05	0.14	−0.04	0.12	−0.23	0.12	−1.16	0.39
Productivity 2000	−0.17**	0.02	−0.09**	0.01	−0.10**	0.013	−0.15**	0.06
Private R&D	0.02**	0.00	0.01	0.01	0.01**	0.00	−0.029	0.03
Public R&D	−0.01	0.01	−0.01	0.04	*−0.01*	*0.00*	*0.04**	*0.01*
Openness economy	−0.03**	0.01	−0.02**	0.01	*−0.04**	*0.01*	*0.04*	*0.04*
Market potential	0.08**	0.01	0.06**	0.01	0.05**	0.01	0.12**	0.03
Education	0.05**	0.01	0.03**	0.01	0.03**	0.01	0.10**	0.05
Population density	0.00	0.00	0.00	0.00	*0.00*	*0.010*	*−0.04**	*0.01*
Wages	0.01	0.01	0.00	0.01	0.01	0.01	0.03	0.05
Related variety	−0.03	0.02	−0.02	0.02	−0.03*	0.02	−0.09	0.07
Unrelated variety	0.00	0.01	0.00	0.01	−0.01	0.01	0.10	0.07
Specialization	0.41**	0.08	0.27**	0.07	*0.19**	*0.06*	*0.96**	*0.26*
W_Productivity growth	0.96**	0.02	0.89	0.04	0.89		0.04	

Summary statistics

	(1)		(2)		(3)			
N	205		205		205			
R^2	0.781		0.837		0.887			
Chow–Wald	–		–		85.8			
BP (heteroskedasticity)	55.01	0	78.453	0	0.033	0.857		
LR (spatial-lag)	57.061	0	113.714	0	127.6	0		
LM (spatial error)	39.49	0	0.051	0.821	0.742	0.389		

Notes: Coefficients and *t*-values. Coefficients that significantly differ over regimes are in italics.
*$p < .10$.
**$p < .05$.

Table 3. Modelling outcomes for unemployment growth 2000–2010

Explanatory variables	(1) Spatial-lag model W_1		(2) Spatial-lag model W_2		(3) Regimes' urban size			
					Small- and medium-sized		Large and capital regions	
(Constant)	1.53	0.84	1.39	0.78	2.51	0.84	−1.12	2.12
Unemployment 2000	−0.45**	0.03	−0.37**	0.03	−0.38**	0.03	−0.07	0.10
Private R&D	−0.07**	0.03	−0.05**	0.02	−0.05*	0.03	0.49**	0.14
Public R&D	0.008	0.026	0.008	0.025	0.031	0.024	−0.124	0.09
Openness economy	0.52**	0.08	0.43**	0.07	0.41**	0.07	1.04**	0.28
Market potential	−0.43**	0.09	−0.32**	0.09	−0.37**	0.09	−0.56*	0.26
Education	−0.13*	0.07	−0.12**	0.06	−0.16	0.06	−0.43	0.35
Population density	0.04	0.03	0.03	0.03	0.01	0.03	0.26**	0.09
Wages	0.12**	0.05	0.08	0.05	0.03	0.05	0.45**	0.17
Related variety	0.21	0.11	0.19	0.10	0.23**	0.10	−0.47	0.48
Unrelated variety	0.05	0.07	−0.010	0.06	0.00	0.06	0.94**	0.44
Specialization	−3.10**	0.53	−2.31**	0.50	−2.95**	0.50	−3.22*	1.79
W_Unemployment growth	0.96**	0.03	0.76**	0.06	0.76**	0.06		
Summary statistics								
N	205		205		205			
R²	0.766		0.814		0.844			
Chow–Wald	–		–		40.6	0		
BP (heteroskedasticity)	24.087	0.012	23.267	0.016	0.016	0.899		
LR (spatial-lag)	77.163	0	101.389	0	107.3	0		
LM (spatial error)	21.683	0	1.992	0.158	5.239	0.022		

Notes: Coefficients and t-values. Coefficients that significantly differ over regimes are in italics.
*p < .10.
**p < .05.

conceptualization in recent years, a pan-European test has until now been missing from the literature. A pan-European test is more interesting than country-level tests, as newly defined cohesion policies, smart-specialization policies, place-based development strategies and policies aimed at fostering competitiveness may be served particularly well by related variety and unrelated variety conceptualizations.

We empirically investigated the contribution of agglomeration economies to economic growth in European regions while separating regions by population size. A conceptual discussion on development burgeons between, on the one hand, spatially blind approaches that argue that intervention regardless of context ("people-based policy") is the best means of development and, on the other hand, place-based approaches that assume that interactions between institutions and geography are more critical for this purpose. This idea has recently been translated into a focus on either the largest regional concentrations ("people-based policies") or an urban network setting combining clusters of cities ("place-based policies"). Our framework combining productivity growth and employment growth shows that spatial regimes classified by the population size of urban regions differ significantly in both sets of models, confirming their *joint* significance. In medium-sized urban regions private R&D and specialization levels (inter alia) are especially important in relation to productivity growth, and sectoral specialization (negatively), related variety and the openness of the economy (inter alia) are especially important in relation to employment growth. In large urban regions, population density (negative), educational level, public R&D and the degree of specialization (inter alia) are relatively more important for productivity growth. The outcomes of these analyses suggest particular roles in development processes for medium-sized ("second-tier") urban regions *alongside* the largest urban regions. Especially related variety—employment growth is a particular feature of small medium-sized urban regions. Perhaps due to agglomeration disadvantages, the largest urban regions do not show the highest employment growth rates. This marked regional heterogeneity indicates that micro-economic processes play out differently in different types of regions, thereby confirming that European place-based policy strategies may play an important role for regional development alongside place-neutral (people-based) policy strategies. However, this heterogeneity also suggests that, similar to European regional innovation patterns, which are differentiated among regions according to their regional context conditions (Camagni & Capello, 2013), regional heterogeneity and inter-regional network positions support the careful consideration of how "smart specialisation" is evaluated in Europe (Thissen *et al.*, 2013).

The hypothesized relationship between unemployment growth and unrelated variety is not confirmed in our first pan-European exercise. This finding suggests that national regulations and institutions in Europe cause the pan-European model to deviate from national models. More research is needed on this issue. In addition, future work should pay more attention to causality (i.e. whether variety induces development or whether developing regions create more variety), panel estimation to ensure the robustness of the relations found, the testing of other types of spatial heterogeneity (e.g. cohesion regions versus core regions, or university regions versus non-university regions), and continuous space modelling of firm-level data to avoid spatial scale and selection processes. Recall that our analyses do not address many of the problems identified in the meta-analyses on measurement and selection issues either. This paper does show that conceptual renewal may represent a fruitful and exciting way to advance the debate on agglomeration and spatial heterogeneity in light of European reforms and policy formulations.

Funding

Research time of Frank van Oort is financed by the Smart Specialization for Innovative Regions (SmartSpec) FP7-project of the European Union (2013).

Notes

1. Studies using the same methodology report similar results in Great Britain (Bishop & Gripaios, 2010; Essletzbichler, 2013), Italy (Antonietti & Cainelli, 2011; Boschma & Iammarino, 2009; Cainelli & Iacobucci, 2012; Mameli *et al.*, 2012; Quatraro, 2010), Germany (Brachert *et al.*, 2011), Finland (Hartog *et al.*, 2012), Spain (Boschma *et al.*, 2012, 2013) and the US (Castaldi *et al.*, 2013).
2. Belgium, Denmark, Finland, France, Ireland, Italy, Portugal, The Netherlands, Spain, Sweden, the UK, Czech Republic, Hungary, Poland, and Slovakia. For other countries data in AMADEUS were found unreliable.

References

Antonietti, R. & Cainelli, G. (2011) The role of spatial agglomeration in a structural model of innovation, productivity and export, *Annals of Regional Science*, 46(3), pp. 577–600.

Barca, F. (2009) *An Agenda for a Reformed Cohesion Policy: A Place-Based Approach to Meeting European Union Challenges and Expectations* (Brussels: Report for the European Commission).

Barca, F., McCann, P. & Rodriguez-Pose, A. (2012) The case for regional development intervention: Place-based versus place-neutral approaches, *Journal of Regional Science*, 52(1), pp. 134–152.

Beaudry, C. & Schiffauerova, A. (2009) Who's right, Marshall or Jacobs? The localization versus urbanization debate, *Research Policy*, 38(2), pp. 318–337.

Bishop, P. & Gripaios, P. (2010) Spatial externalities, relatedness and sector employment growth in Great Britain, *Regional Studies*, 44(4), pp. 443–454.

Boschma, R. A. (2005) Proximity and innovation: A critical assessment, *Regional Studies*, 39, pp. 61–74.

Boschma, R. A. & Iammarino, S. (2009) Related variety, trade linkages, and regional growth in Italy, *Economic Geography*, 85(3), pp. 289–311.

Boschma, R. A. & Lambooy, J. G. (1999) Evolutionary economics and economic geography, *Journal of Evolutionary Economics*, 9(4), pp. 411–429.

Boschma, R. A. & Martin, R. (2010) *The Handbook of Evolutionary Economic Geography* (Cheltenham: Edward Elgar).

Boschma, R., Minondo, A. & Navarro, M. (2012) Related variety and economic growth in Spain, *Papers in Regional Science*, 91(2), pp. 241–256.

Boschma, R., Minondo, A. & Navarro, M. (2013) The emergence of new industries at the regional level in Spain. A proximity approach based on product-relatedness, *Economic Geography*, 89(1), pp. 29–51.

Brachert, M., Kubis, A. & Titze, M. (2011) *Related variety, unrelated variety and regional functions: Identifying sources of regional employment growth in Germany from 2003 to 2008*. IWH-Diskussionspapiere no. 2011, p. 15.

Cainelli, G. & Iacobucci, D. (2012) Agglomeration, related variety, and vertical integration, *Economic Geography*, 88(3), pp. 255–277.

Camagni, R. & Capello, R. (2013) Regional innovation patterns and the EU regional policy reform: Towards smart innovation policies, *Growth and Change*, 44(2), pp. 355–389.

Castaldi, C., Frenken, K. & Los, B. (2013) Related variety, unrelated variety and technological breakthroughs: An analysis of U.S. state-level patenting. Papers in Evolutionary Economic Geography 13.02, Utrecht University, Utrecht.

Combes, P. P. & Overman, H. G. (2004) The spatial distribution of economic activites in the European Union, in: J. V. Henderson & J. F. Thisse (Eds) *Handbook of Regional and Urban Economics*, pp. 2846–2909 (Dordrecht: Elsevier).

De Groot, H. F. L., Poot, J. & Smit, M. J. (2009) Agglomeration externalities, innovation and regional growth: Theoretical reflections and meta-analysis, in: R. Capello & P. Nijkamp (Eds) *Handbook of Regional Growth and Development Theories*, pp. 256–281 (Cheltenham: Edward Elgar).

Desrochers, P. & Leppälä, S. (2011) Opening up the 'Jacobs Spillovers' black box: Local diversity, creativity and the processes underlying new combinations, *Journal of Economic Geography*, 11(5), pp. 843–863.

Dijkstra, L., Garcilazo, E. & McCann, P. (2013) The economic performance of European cities and city regions: Myths and realities, *European Planning Studies*, 21(3), pp. 334–354.

Dogaru, T., van Oort, F. G. & Thissen, M. (2011) Agglomeration economies in European regions: Perspectives for objective-1 regions, *Journal of Economic and Social Geography (TESG)*, 102(4), pp. 486–494.

Duranton, G. & Puga, D. (2000) Diversity and specialisation in cities: Why, where and when does it matter? *Urban Studies*, 37(3), pp. 533–555.

Essletzbichler, J. (2013) Relatedness, industrial branching and technological cohesion in US metropolitan areas, *Regional Studies*. doi: 10.1080/00343404.2013.806793

ESPON. (2012) *Second Tier Cities and Territorial Development in Europe: Performance, Policies and Prospects (SGPTD)* (Brussels: ESPON).

Frenken, K., van Oort, F. G. & Verburg, T. (2007) Related variety, unrelated variety and regional economic growth, *Regional Studies*, 41(5), pp. 685–697.

Glaeser, E. L., Kallal, H., Scheinkman, J. & Shleifer, A. (1992) Growth in cities, *Journal of Political Economy*, 100(6), pp. 1126–1152.

Hartog, M., Boschma, R. & Sotarauta, M. (2012) The impact of related variety on regional employment growth in Finland 1993–2006: High-tech versus medium/low-tech, *Industry and Innovation*, 19(6), pp. 459–476.

Henderson, V., Kuncoro, A. & Turner, M. (1995) Industrial development in cities, *Journal of Political Economy*, 103(5), pp. 1067–1085.

Jacobs, J. (1969) *The Economy of Cities* (New York: Random House).

Le Gallo, J. & Kamarianakis, Y. (2011) The evolution of regional productivity disparities in the European Union from 1975 to 2002: A combination of shift-share and spatial econometrics, *Regional Studies*, 45(1), pp. 123–139.

Mameli, F., Iammarino, S. & Boschma, R. (2012) Regional variety and employment growth in Italian labour market areas: Services versus manufacturing industries. Papers in Evolutionary Economic Geography 12.03, Utrecht University, Utrecht.

Marshall, A. (1890) *Principles of Economics* (London: MacMillan).

McCann, P. & van Oort, F. G. (2009) Theories of agglomeration and regional growth: A historical review, in: R. Capello & P. Nijkamp (Eds) *Handbook of Regional Growth and Development Theories*, pp. 19–32 (Cheltenham: Edward Elgar).

Melo, P. C., Graham, D. J. & Noland, R. B. (2009) A meta-analysis of estimates of agglomeration economies, *Regional Science and Urban Economics*, 39(3), pp. 332–342.

Moreno, R., Paci, R. & Usai, S. (2005) Spatial spillovers and innovation activity in European regions, *Environment & Planning A*, 37(10), pp. 1793–1812.

Neffke, F., Henning, M. & Boschma, R. (2011) How do regions diversify over time? Industry relatedness and the development of new growth paths in regions, *Economic Geography*, 87(3), pp. 237–265.

OECD. (2009) *Regions Matter: Economic Recovery, Innovation and Sustainable Growth* (Paris: OECD).

OECD. (2011) *Regional Outlook. Building Resilient Regions for Stronger Economies* (Paris: OECD).

OECD. (2012) *Promoting Growth in All Regions* (Paris: OECD).

OECD. (2013) *Redefining "Urban". A New Way to Measure Metropolitan Areas* (Paris: OECD).

O'Huallachain, B. & Lee, D. S. (2011) Technological specialization and variety in urban invention, *Regional Studies*, 45(1), pp. 67–88.

Paci, R. & Usai, S. (1999) Externalities, knowledge spillovers and the spatial distribution of innovation, *Geojournal*, 49(4), pp. 381–390.

Puga, D. (2002) European regional policies in light of recent location theories, *Journal of Economic Geography*, 2(4), pp. 373–406.

Quatraro, F. (2010) Knowledge coherence, variety and economic growth: Manufacturing evidence from Italian regions, *Research Policy*, 39(10), pp. 1289–1302.

Rosenthal, S. S. & Strange, W. C. (2004) Evidence on the nature and sources of agglomeration economies, in: J. V. Henderson & J. F. Thisse (Eds) *Handbook of Regional and Urban Economics*, pp. 2119–2171 (Amsterdam: Elsevier).

Shefer, D. & Frenkel, A. (1998) Local milieu and innovativeness: Some empirical results, *The Annals of Regional Science*, 32(1), pp. 185–200.

Thissen, M., van Oort, F., Diodato, D. & Ruijs, A. (2013) *Regional Competitiveness and Smart Specialization in Europe. Place-Based Development in International Economic Networks* (Cheltenham: Edward Elgar).

Van Oort, F. G. (2004) *Urban Growth and Innovation. Localized Externalities in the Netherlands* (Aldershot: Ashgate).

Van Oort, F. G. & Bosma, N. (2013) Agglomeration economies, inventors and entrepreneurs as engines of European regional development, *Annals of Regional Science*, 51(1), pp. 213–244.

Van Oort, F. G. & Lambooy, J. G. (2014) Cities, knowledge and innovation, in: M. M. Fischer & P. Nijkamp (Eds) *Handbook of Regional Science*, pp. 475–488 (Berlin: Springer).

World Bank. (2009) *World Development Report: Reshaping Economic Geography* (Washington, DC: World Bank).

Assessing Polycentric Urban Systems in the OECD: Country, Regional and Metropolitan Perspectives

MONICA BREZZI & PAOLO VENERI

Organisation for Economic Co-operation and Development, Regional Development Policy Division, GOV/RDP, Paris, France

ABSTRACT   *Contemporary urban systems in OECD countries are structured around functional regions, which often overcome established city-boundaries. Reading space in terms of functional regions allows assessing changes in urban hierarchies and spatial structures, including the polycentricity of urban systems at national, regional and metropolitan scales. By using a harmonized definition of functional urban areas in OECD countries, this paper first provides a sound definition of polycentricity at each spatial scale, highlighting for each of them the different links with policy. Second, it provides measures of polycentricity and explores the economic implications of different spatial structures. Results show that relatively more monocentric regions have higher GDP per capita than their more polycentric counterparts. At the country level, on the other hand, polycentricity is associated with higher GDP per capita.*

1. Introduction

The way people and economic activities organize in space has been changing over time. Improvements in communication technologies, large and growing movements of people and goods, and economic development processes have generated an enlargement of the spaces where people live and work. These changes facilitated suburbanization processes and an increasing integration of cities with their surrounding hinterland. The emerging spaces where people live and work and where the bulk of economic interdependencies takes place is referred to in the literature as "functional regions". Among functional regions, the "functional urban areas" (FUAs) are characterized by the presence of one or more urban centres of different sizes and economic importance.

A better knowledge of urban spatial organization can have important consequences on regional policy making. Half of OECD population live in the 275 large FUAs (metropolitan areas) that contribute to more than 50% of GDP and employment of the entire OECD. While the concentration of people in dense urban centres of "established" OECD cities has slowed down or even decreased in some cases, the coming together of people and business in urban areas of varying sizes has not stopped. Moreover, the reduction of transport and communication costs will continue to make urban centres increasingly interconnected and change urban areas from monocentric agglomerations to a more "polycentric" system of integrated urban centres and sub-centres. The way people in cities have access to education and jobs, decent housing, efficient transportation, and a safe and sustainable environment will have a strong impact on national and global prosperity, and thus a better understanding of the different urbanization forms will help recognize the impact of different national and local strategies.

Polycentric agglomerations can be investigated at higher territorial scales beyond metropolitan areas. For example, in many regions a number of cities and towns are increasingly linking up and, similarly, OECD countries differ in the spatial organization and connections among urban areas. Understanding the functioning and efficiency of these connections can help clarify the links between urbanization and economic development. Metropolitan areas tend to be more productive than other regions, and on average the GDP per capita in OECD metropolitan areas was 15% higher than in the rest of the economy in 2010 (OECD, 2013a).

Previous works have investigated the role of polycentricity in modern urban systems both for specific countries and comparatively for European countries (Espon, 2003; Vandermotten *et al.*, 2007). The policy relevance of understanding the links between the spatial organization of urban areas and the socio-economic conditions of a country or a region is apparent: both for national and local policy makers, it can help targeting policies, planning public services and designing the institutional organization and governance mechanisms that can best support the development of different territories and contribute to national growth. However, the empirical evidence on the links between polycentricity and economic development is not conclusive and results seem to depend on the choice of countries, the conceptual definition of polycentricity and on the indicators chosen to measure this characteristic of spatial structure.

This paper assesses the polycentric structure of OECD urban systems at three different geographic scales: in a metropolitan area, a region and in a country. It does so by relying on a common definition of FUAs applied to 29 OECD countries (OECD, 2012a). Secondly, the paper explores the implication of polycentricity at national and regional level on the overall level of economic prosperity. The results show that OECD countries with a more polycentric urban structure are associated with higher GDP per capita. However, at the regional level, relatively more monocentric regions have higher GDP per capita than their more polycentric counterparts.

The paper is organized as follows. Section 2 briefly recalls the method to define OECD FUAs. Then, it articulates the concept of polycentricity at three different scales: in a metropolitan area, as a network of urban areas in a larger region, and as the urban system of a country. Section 3 provides some evidence on the spatial organization of the OECD metropolitan areas and recent changes towards sprawl or instead more compact development. Sections 4 and 5 apply the definition of polycentricity at regional and country levels, respectively. In both cases a preliminary analysis of the links between polycentricity and levels of GDP per capita is carried out. Possible improvements of these results, including building internationally comparable measures of polycentricity based on the functions carried out by cities, are discussed in Section 6, which concludes.

2. Polycentric Development and FUAs

2.1. *Overview of the Methodology to Identify FUAs in OECD Countries*

The method to identify FUAs within OECD countries integrates geographic information sources (GIS) with administrative and survey sources to capture the highly densely populated areas (urban cores) and the commuting flows towards the urban cores regardless of the administrative boundaries (Figure 1).

The method consists of three main steps: the first step identifies contiguous or highly interconnected densely inhabited urban cores, by using population grid data at 1 km^2. The second step of the procedure allows the identification of urban cores that are not contiguous but belong to the same FUA. Two urban cores are considered integrated, and thus part of the same polycentric metropolitan area, if more than 15% of the working population of any of the cores commutes to work in the other core. The third step defines the commuting shed or hinterland of the FUA, by selecting those municipalities that send at least 15% of their work force to the cores. More details can be found in OECD (2012a).

2.2. *Defining Polycentricity*

The first necessary step of this analysis is a clear definition of the concept of polycentricity. Starting with the most general meaning, any given area can be defined as polycentric if it contains two or more centres. With just a bit more precision, an area is polycentric if its population or employment is not concentrated to a substantial extent in one single centre (Riguelle *et al.*, 2007, p. 195). Considering the distribution of population and employment in space means to interpret polycentricity as a "morphological" concept. It should be pointed out that the literature distinguishes between a morphological

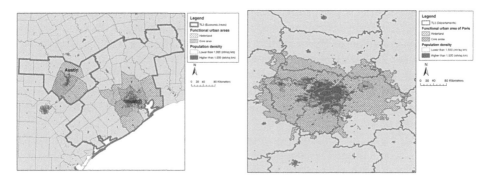

Figure 1. Urban and non-urban population density; functional and administrative boundaries: Houston and Paris.
Source: OECD calculations based on population density disaggregated with Corine Land Cover, Joint Research Centre for the European Environmental Agency.

dimension—which focuses on population, employment, land use, etc.—and a more "functional" one (Nordregio, 2005; Burger and Meijers, 2012; Veneri, 2013a)—linked to the functions carried out by cities or the connections among them (e.g. commuting flows). However, the two dimensions are very much related to one another. The morphological dimension of polycentricity focuses on the size and distribution of urban centres across space. This dimension is often associated with the extent to which territory is characterized by a balanced development. The functional dimension of polycentricity focuses less on the internal characteristics of the centres—such as size, density, etc.—and more on the way these centres organize the rest of the territory by supplying the functions that shape the territorial hierarchies (Green, 2007; De Goei et al., 2010).

Whether a functional or a morphological approach to polycentricity is adopted can also depend on the territorial level at which this concept is applied. For the sake of simplicity, ESPON 3.1 (2003) classified the territorial scales in three categories, namely "micro", "meso" and "macro" ones. Departing from the Espon classification, this paper considers three major perspectives with which to look at polycentricity, namely the metropolitan, the regional and the national perspectives. Adopting the former scale implies considering the spatial organization within the metropolitan space, which is a space characterized by one single—or multiple overlapping—labour market areas. On the other hand, the national perspective looks at the spatial structure of the entire national urban system or, in the case of the European space, supra-national urban system. Finally, the regional perspective refers to networks of two or more FUAs which are connected through functional relationships and lie in the same larger administrative region. This intermediate perspective draws on concepts that have been widely analysed in the literature, such as the polycentric urban region (PUR) (Dieleman & Faludi, 1998; Kloosterman & Musterd, 2001; Parr, 2004). Different spatial scales are associated with different meanings and potential policy issues at stake. For urban systems at each scale, metropolitan, regional and national, the polycentric spatial structure is interpreted and analysed through the lens of the FUAs, as identified by the OECD. Table 1 proposes a summary.

Table 1. Polycentricity at three spatial scales: a summary

Geographical scale	Measures	Potential policy issues
Intra-metropolitan	Sprawl Index	Efficiency in land use
	Share of people and jobs in urban centres	Environmental concerns
		Transport and public services efficiency
Regional (inter-metropolitan)	Relative importance of the largest city	Regional agglomeration economies
	Size distribution of cities	Intra-regional territorial disparities in access to services and variety of consumption
	Connectivity among cities	
National	Relative importance of the largest city	Possible need for a national policy for urban areas to focus on the potential of all cities, fostering agglomeration economies and ensuring policy coherence.
	Size distribution of cities	Territorial disparities (income, services, consumption)

3. Metropolitan Scale

Spatial structure at the metropolitan scale has a multi-dimensional policy relevance. The way population and economic activities distribute across the metropolitan space can affect the economic performance of metropolitan areas, through shaping the intensity of agglomeration economies (García-López & Muñiz, 2013). In addition, spatial structure can be important for efficiency in the provision of public services. Public transport can be more efficiently organized when people and jobs are concentrated in centres of a certain size, which ensure the achievement of economies of scale. Other issues regard energy consumption, green space and land use. Regarding transport, for example, the degree of metropolitan polycentricity has been found to be associated with higher car dependency (Glaeser & Kahn, 2004), but evidence on the possible effect on travel time and distance is still ambiguous (Schwanen *et al.*, 2004; Veneri, 2010; Modarres, 2011).

Assessing polycentricity at the metropolitan scale means to put the focus on a self-organized and economically integrated space, often characterized by a single labour market area or several overlapping ones. At this scale, spatial structure has been traditionally conceptualized in urban economics as monocentric, with a central business district (CBD) located at the centre of the area (Alonso, 1964; Muth, 1969; Mills, 1972). The CBD is characterized by the highest job density, which declines monotonically as the distance from the CBD increases. However, metropolitan areas have been expanding in the last decades and their spheres of influence have regionalized. Their extension goes often well beyond traditional administrative boundaries and, as a consequence, other new or pre-existing centres coalesce or integrate in the larger "functional region" (Champion, 2001) or emerge from a decentralization process from the CBD (Anas *et al.*, 1998). These processes challenged traditional monocentric models in urban economics and stimulated the introduction of new models, which incorporate the possibility of polycentric and dispersed structures (Anas *et al.*, 1998; White, 1999).

Figure 2 shows the density patterns of population in the metropolitan areas of Paris and San Francisco. The figure shows that despite density decreasing, on average, as distance to the main centre increases, this pattern is not monotonic and there are several local peaks of

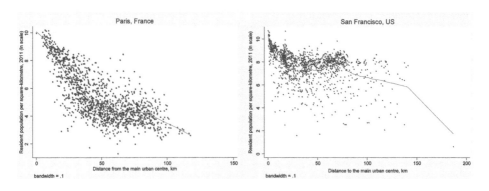

Figure 2. Density patterns in the FUAs of Paris (France) and San Francisco (USA). Density does not decrease monotonically as distance to the main centre increases.
Note: Units of analysis are municipalities in the case of Paris and Census tracks for San Francisco.
Source: Author's elaboration on National Census data.

high density. This should indicate the presence of metropolitan sub-centres, hence a poly-centric spatial structure. Under this perspective, one simple way to measure the degree of metropolitan polycentricity consists in identifying those spatial units that can be considered as sub-centres and computing their share of population (or employment) over total metropolitan population (or employment).

Metropolitan polycentricity can also be seen from a different angle, by focusing on morphological features and land development patterns. Under this perspective, metropolitan polycentricity can be seen as a model of urban development that is an alternative to dispersion and that is sometimes called "decentralised concentration" (Frey, 1999). In principle, it combines the need to accommodate urbanization with that of limiting generalized dispersion of activities across space, which is often referred to as sprawl or "Edge-less" city (Lang, 2003; OECD, 2012b). Using the OECD definition of FUAs, it is possible to look at the dynamics of land use and to assess whether metropolitan areas are following patterns of development towards sprawl or compactness. According to Brueckner (2001), sprawl is defined here as the "excessive" urbanization. A simple Sprawl Index (SI) has been developed to measure the growth in built-up area adjusted for the growth in city population (OECD, 2013a). When the population is stable, the SI corresponds to the growth of the built-up area. When the city population changes, the index measures the increase in the built-up area relative to a benchmark where the built-up area would have increased in line with population growth. The index is calculated as in the formula below (1):

$$SI_i = \frac{[urb_{i,t+n} - (urb_{i,t}*(pop_{i,t+n}/pop_{i,t}))]}{urb_{i,t}}*100, \tag{1}$$

where i refers to the ith metropolitan area; t refers to the initial year; $t + n$ is to the final year; urb is to the number of square kilometres of the total built-up area; pop is total population.

The Sprawl Index shows a high degree of heterogeneity between the patterns of urban development within metropolitan areas in Europe, Japan and the USA.[1] There is not an overall sprawling pattern emerging. Between 2000 and 2006 one-third of the metropolitan areas experienced positive variation in the Sprawl Index (with the average value for the OECD being 0.8%), hence the growth of built-up land was faster than the growth of the population. In other words the built-up area per person has increased. Figure 3 shows some of the metropolitan areas with the highest values of the Sprawl Index. Several metropolitan areas of Japan, Las Palmas and Zaragoza (Spain) and Tallin (Estonia) show values higher than 10%. However, they had relatively lower levels of built-up area per person in 2000, compared to metropolitan areas in the USA.

The patterns of spatial structure within metropolitan areas can have different implications in economic, environmental and social terms. While this section does not explore empirically any of these relationships, it is enough to say that much work has been produced on the costs of sprawl, despite the lack of agreement on the actual rationality of policies aimed at containing this phenomenon. On policentricity, on the other hand, much less work has been done to understand whether this pattern of spatial development might improve the efficiency in land use or the environmental conditions if compared to sprawl.

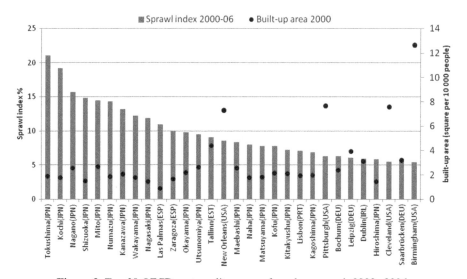

Figure 3. Top 30 OECD metropolitan areas for urban sprawl, 2000–2006.
Source: Own calculations from OECD metropolitan database. Only Europe, Japan and the USA are included.

4. Network of Cities: Polycentricity at Regional Level

New concepts have been introduced in the last two decades to identify and describe regional spatial structures where several urban areas co-exist and might be able to generate positive externalities beyond the boundaries of the urban areas. Among these concepts, Dieleman and Faludi (1998) refer to "Polynucleated Urban Field", while Parr uses the expression PUR (Parr, 2004). Following Parr, PURs are regions—which can be both administratively defined or approached functionally—that are organized around several urban areas. These areas should be morphologically separated but still kept in close physical proximity and, in order to develop synergy, they should be functionally connected and/or have complementary sectoral specializations (Parr, 2004, p. 232). A PUR is hence characterized by a substantial equilibrium—or low hierarchy—among cities in terms of population and economic power (Bailey & Turok, 2001) and from a clear physical separation between its centres.

The characteristics of regional spatial structures can have different implications in terms of economic outcomes through, for example, agglomeration economies (Duranton & Puga, 2004; Rosenthal & Strange, 2004; Combes *et al.*, 2012) or consumption benefits, ensured by a higher variety of consumption possibilities in large agglomerations (Glaeser *et al.*, 2001). These advantages can be reached in large cities with high population and job density. However, it has also been argued that the advantages of agglomeration can be "regionalized", and achieved in regions characterized by the presence of several interconnected urban centres. This hypothesis was first advanced by Alonso (1973) through the idea that polycentric regions can exploit synergies emerging from co-operation and complementarities among centres, as if each centre could "borrow" its size from the rest of the region, compensating for an eventual lack of large agglomerations.

The potential economic implications of polycentric spatial structures include, in addition to the economies of scale (borrowing size), also other aspects linked to the complexity and diversity of functions. In the words of Parr (2002), these advantages should be named "regional externalities", considered less strict in terms of spatial agglomeration of individuals and organizations than agglomeration and localization externalities. Regional externalities can emerge especially in the context of regions characterized by city-network relationships, providing benefits such as sharing high-scale infrastructures, highly qualified specialized services (Priemus, 1994) or exploiting regional complementarities that might emerge from local endowments of hard and soft factors of development.

Using the OECD definition of FUAs, it is possible to identify the nodes of regional polycentricity in OECD large administrative regions (TL2). The latter are chosen as units of analysis, since they represent the first tier of sub-national government in OECD countries and are sufficiently comparable in terms of functions. Figure 4 shows two TL2 regions in Europe of similar area size, namely Aragon (Spain) and Brittany (France), which appear to have very different spatial structures. The former is characterized by one single major metropolitan area only (Zaragoza), while in the latter region several small and medium-sized FUAs co-exist with the metropolitan area of Rennes. Close-by FUAs—especially when sharing a common higher administrative tier—are often involved in co-operation initiatives for purposes of economic development strategy or service provision and recent OECD analyses provide several examples in this respect (OECD, 2013b).

Through a consistent delineation of FUAs, regional polycentricity across 29 OECD countries is assessed quantitatively in this section, which then explores how polycentricity is associated with the level of regional economic development. The first step consists in identifying a measure of polycentric development. In principle, a complete and sound measure of polycentricity should take into account the population size of centres, their distribution and their connectivity (Wegener, 2013). Considering all these dimensions at the same time helps not to lose too much of the complexity of polycentric spatial structures and to adopt at the same time some morphological (size and distribution) and functional

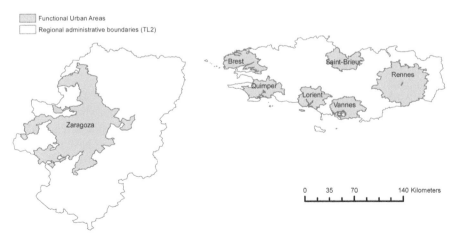

Figure 4. Monocentric vs. polycentric regional spatial structures: the case of Aragon (Spain) and Brittany (France).
Source: Authors' elaborations from OECD (2012a).

95

(connectivity) features. However, this may imply the use of composite indicators, which are more discretional and require a much higher amount of data. For the purpose of this paper, simple and straightforward measures of polycentricity for regions and countries have been preferred, following existing literature (Meijers & Sandberg, 2008; Veneri & Burgalassi, 2012). The first is urban primacy, which considers the share of population in the most populated city over the total regional population or over the sum of the urban population of a region. For the OECD TL2 regions, urban primacy is here defined as the share of population in the most populated FUA over total regional population. Intuitively, the higher the primacy, the higher the regional monocentricity.

Polycentricity at the regional (and national) scale can also be measured through the beta coefficient of the following equation:

$$\ln (\text{rank}) = \alpha + \beta \ln (\text{size}), \tag{2}$$

where size is total population of each FUA within a given region; rank is the rank, computed by region, of FUAs by size. The slope of the line interpolating data, given by the estimated beta, indicates the level of hierarchy among FUAs, and thus the level of polycentricity of each region. By definition, the beta coefficient is negative. In absolute terms, the higher the value of beta—hence the steeper the line interpolating data—the higher the level of polycentricity. The use of functionally defined urban areas as building blocks for regional polycentricity allows beta coefficients to approximate the hierarchical distribution of cities over regional territory without making the mistake of considering places that are part of a single integrated area (e.g. municipalities) as separated nodes of the urban systems (Kloosterman & Musterd, 2001; Parr, 2004).

Matching FUAs with TL2 administrative boundaries is not always straightforward. Some large FUAs (e.g. Paris, Prague, etc.) cover a space that is larger than the administrative region where they are located. In some other cases, several FUAs fall within the boundaries of one single region. In order to minimize possible inconsistencies, FUAs are allocated to regions on the basis of the location of the urban core only. However, all indicators of spatial structure, such as primacy and polycentricity, are computed by considering data at the level of the whole FUAs (both cores and hinterlands). Adopting this method, there are still regions with a value of urban primacy that is larger than 1. This can happen when FUAs' cores cross different administrative regions. In order to avoid biases—and for the sake of simplicity—these regions have not been considered in the analysis.

In this work, the coefficients of equation (2) were estimated for OECD TL2 regions in order to assess their degree of polycentricity. Clearly, not all the regions have FUAs inside their territories. Some have no FUAs, hence they can be classified as rural regions, while others have just one, hence they are considered as monocentric. Others have two or more, so they are polycentric. Consistently with Meijers and Burger (2010) coefficients of equation (2) were estimated taking into account, for each region, the four largest FUAs only, so as to ensure consistency in the number of observations considered and a consequent higher comparability among regions. Figure 5 shows the rank-size distribution of the four main FUAs in the Capital region of Korea and Brittany (France). A steeper slope of the line interpolating data indicates a higher degree of polycentricity. The Korean Capital Region is the one with the lowest level of polycentricity among 147 OECD regions with at least four FUAs within their respective territory.

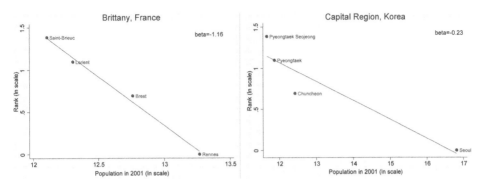

Figure 5. Rank-size distribution of FUAs in Brittany (France) and Capital Region (Korea). *Source*: Authors' elaboration on OECD metropolitan database.

Primacy and polycentricity represent two key features of regional spatial structure. A regression analysis was carried out in order to provide some first international evidence on how regions characterized by a presence of several cities of similar size—hence those that are more polycentric—are associated with socio-economic conditions. According to the hypothesis of "borrowing size" the advantages emerging from large agglomerations should be compensated by a polycentric spatial structure. More specifically, this relationship was explored by looking at how the levels of regional GDP per capita are associated with characteristics of spatial structure, after controlling for other basic factors.

Data used in the analysis come from the OECD regional and metropolitan database. The dependent variable is the natural logarithm of GDP per capita expressed in Purchasing Power Parity US$. Independent variables included in the analysis are the following: "size" is the total regional population; "education" is the share of workforce holding tertiary education; "primacy" is the share of population in the largest FUA located in the region over total regional population; "d_polycentricity" is a dummy equal to 1 when the region has at least two FUAs in its territory, 0 otherwise; "beta*polycentricity" is a direct measure of the degree of regional polycentricity and it is computed as the interaction between "d_polycentricity" and the beta coefficients of the rank-size equation (2); "metro_share" is the share of regional population that live in FUAs; "d_East_Europe" and "d_West_Europe" are dummies equal to 1 when the region is located in Eastern European countries or Western European countries, respectively.[2] These dummies were included in order to control for the spatial heterogeneity in OECD countries and for some peculiar characteristics of Europe related with both spatial development and policy. In fact, the idea of preserving the polycentric development that historically characterizes European territory has been shaping the EU spatial policy (Davoudi, 2003). All the variables refer to 2010.

Table 2 reports the results of different model specifications. Model 1 considers primacy only (share of regional population living in the largest city) as feature of spatial structure, while Model 2 considers polycentricity. Model 3 includes both primacy and polycentricity, while Model 4 controls for the total share of people living in FUAs instead of the primacy. These different specifications help interpreting results with more robustness, given possible collinearity among different variables of spatial structures ("primacy" and "beta*polycentricity" show a linear correlation of -0.39).[3] Results confirm that

Table 2. OLS estimation results

Variable	Model 1		Model 2		Model 3		Model 4	
constant	−5.933	(0.432)***	−6.271	(0.509)***	−6.424	(0.506)***	−6.245	(0.518)***
Size	0.085	(0.028)***	0.122	(0.038)***	0.128	(0.037)***	0.121	(0.038)***
education	2.420	(0.345)***	2.546	(0.341)***	2.399	(0.341)***	2.559	(0.345)***
d_East_Europe	−0.037	(0.093)	0.066	(0.098)	0.059	(0.097)	0.062	(0.099)
d_West_Europe	0.354	(0.074)***	0.423	(0.073)***	0.418	(0.073)***	0.422	(0.074)***
primacy	0.372	(0.134)***			0.256	(0.137)*		
d_polycentricity			−0.055	(0.113)	−0.052	(0.112)	−0.167	(0.061)***
beta*polycentricity			−0.170	(0.057)***	−0.149	(0.058)**	−0.048	(0.114)
metro_share							−0.017	(0.075)
Number of obs.	206		206		206		206	
Adj. R-squared	0.376		0.395		0.402		0.393	
mean VIF	1.17		1.39		1.38		1.40	

Notes: Dependent variable: GDP per capita in 2010 (US$, PPP). Robust standard errors are reported in parentheses.
*Statistically significant at 10% confidence level.
**Statistically significant at 5% confidence level.
***Statistically significant at 1% confidence level.

Source: Authors' elaboration on OECD regional and metropolitan database.

once controlling for overall size, education levels and macro-geographical location, regions where a larger part of the population is located in the largest FUA (higher primacy) have, on average, higher GDP per capita. This result is consistent with previous analyses focusing on single countries or in Europe only (Cervero, 2001; Vandermotten *et al.*, 2007; Veneri & Burgalassi, 2012).

Regarding polycentricity, regions characterized by a lower degree of polycentricity—hence those with a more hierarchical system—are associated with higher levels of GDP per capita, consistently with other studies focusing on single countries (Veneri & Burgalassi, 2011, 2012). The extent to which one single or more FUAs are located within regional territories emerges instead as not correlated with the dependent variable ("d_polycentricity" is not statistically significant). The sign of the coefficient related to the "beta*polycentricty" variable does not support the hypothesis that higher levels of GDP per capita are correlated to a polycentric structure of urban centres of different sizes (i.e. that smaller urban centres can functioning like larger ones by "borrowing size" from the larger ones in the same region). Regarding the other variables included in the analysis, regional size and the share of people with high education are associated with higher levels of GDP per capita, consistently with expectations and with existing literature. The total share of urban population, instead, does not emerge to be statistically significant.

These results should be seen as a preliminary exploration of a complex and multi-faceted relationship between spatial structure and regional socio-economic conditions from an international comparative perspective. The need to use different measures of polycentricity that can account also for the connectivity among urban centres should be taken into account. In addition, linking the self-organization of the territory (FUAs) with the administrative structure (TL2 regions), is not obvious, since, by definition, functional territories can easily cross regional administrative boundaries. This makes it more difficult to allocate FUAs to the "right" administrative entity.

5. Polycentric Development at Country Level

Assessing the polycentric structure of national urban systems is a relevant task for policy. Especially in the current economic downturn that many OECD countries have been facing in the last few years, there is a need to understand what contribution to national prosperity comes from the different regions and whether investment priorities should move towards few large and capital cities or spread in a wider set of cities (Dijkstra *et al.*, 2013; Parkinson *et al.*, 2014). In Europe, a polycentric spatial development is seen as a tool to ensure a more balanced, competitive and sustainable territorial development (ESDP, 1999). In this respect, a research question emerging from this policy framework concerns the role of polycentricity as a way to achieve better development and lower inequalities at the country level.

The spatial structure of national urban systems has been studied for a long time. Overall, the empirical evidence indicates that the distribution of cities over space follows a power law—mostly in the form of a Pareto distribution—meaning that the product of the rank and the size of cities is a constant (Cheshire, 1999; Gabaix & Ioannides, 2004; Veneri, 2013b). Several theoretical explanations have been provided for this empirical evidence, from random shocks related with population migration, productivity and innovation (Gabaix, 1999; Eeckhout, 2004; Duranton, 2007) to Christallerian approaches based on functions played by cities of different sizes (Hsu, 2012). Notwithstanding the robust evidence on the regularity in the relationship between rank and size of cities at the national level, spatial structures of urban systems differ across countries. Some countries are more polycentric than others, meaning that they are organized around a flatter urban hierarchy, where the latter is reflected by the coexistence of more cities of similar size, especially in the right tail of the size distribution.

Consistently with the interpretation carried out at the regional level, the rank–size relationship of the FUAs located in a country gives an idea of the relative importance of large and small cities as well as a measure of the degree of national polycentricity. Higher values of beta coefficients (in absolute value) from the rank-size estimation indicate a higher degree of national polycentricity. Figure 6 plots the rank-size relationship in the natural logarithm scale for the national urban systems of Korea and Germany. The higher absolute value of the beta coefficient for Germany indicates a higher degree of polycentricity compared with the Korean urban system.

In order to measure and compare the degree of polycentricity of national urban systems, beta coefficients were estimated using, for each country, the four largest FUAs only, consistently with what was previously done at regional level. This choice allows comparing 26 OECD countries, excluding only Luxembourg, Estonia and Slovenia, which have 1, 3 and 2 FUAs, respectively. Focusing on the four largest FUAs is also a way to better account for the differences in size—and hierarchical rank—among the largest cities, also catching some aspects of urban primacy.

At the national level, the degree of polycentricity appears to be positively correlated with average levels of economic prosperity in the 26 OECD countries considered. Figure 7 (partial residual plot) shows that more polycentric countries show on average higher levels of GDP per capita. The figure plots the relationship between the degree of polycentricity (*beta* coefficients at national level) and the natural logarithm of GDP per capita expressed in Purchasing Power Parity US$ in 2010 after having controlled for few other basic variables[4] in a simple linear model. Controls include the degree of urban primacy, the share of workforce holding tertiary education (*education*)—all referred to 2010—and two dummies equal to 1 when the country is located in Eastern Europe or Western Europe, respectively. While it must be acknowledged that the limited number of observations and the cross-country nature of the regression analysis make only a

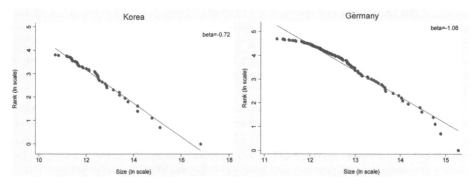

Figure 6. Rank-size distributions of FUAs in Korea and Germany.
Source: Authors' elaboration on OECD metropolitan database.

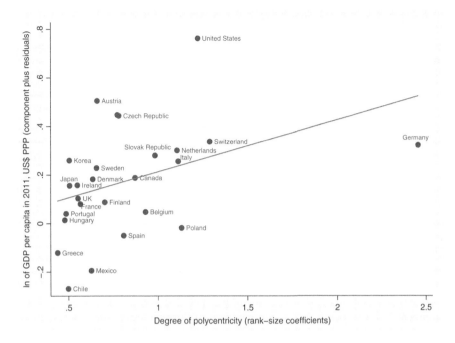

Figure 7. Degree of polycentricity and GDP per capita at national level, 2010 (partial residual plot). *Source*: Authors' elaborations on OECD data (http://dotstat.oecd.org/Index.aspx).

first description of the relationship under investigation possible, it is interesting to note that this result is opposite to the one emerging at the regional level.

The relationship plotted in Figure 7 suggests that polycentricity at country level is not likely to reflect the borrowing size mechanism that was hypothesized at the regional scale. In fact, urban and metropolitan areas in polycentric countries are not necessarily located in close proximity, and network relationship might be less important. The potential advantages of polycentricity at the country scale in terms of economic conditions may come from lower levels of agglomeration costs, which can be spread throughout several FUAs. Another possible explanation is that in a more polycentric structure, a bigger part of the national territory benefits from being close to at least one large FUA compared to, for example, a situation where people are more concentrated in just one large FUA. As far as urban primacy is concerned, this turns out to be not associated with the levels of GDP per capita.

6. Concluding Remarks

This work contributes to the understanding of the spatial structure of urban systems and its relevance for policy. It focuses on polycentricity, which is defined and treated separately at three different spatial scales: metropolitan, regional and national. The conceptual definition of polycentricity as well as the indicators used to measure such phenomena builds on a consistent definition of FUAs across OECD countries. FUAs are used as building block to assess, on a wide international basis, the polycentric spatial structure of metro-politan, regional and national urban systems.

The possibility of comparing a large set of different OECD countries imposed a high approximation in the measures adopted to assess polycentricity. It has to be specified that this work considers polycentricity in its morphological dimension, not taking into account the connectivity among centres, which is another key dimension of this phenom-enon, especially at the regional level. In this respect, two possible refinements of the analy-sis could be considered. First, further effort is needed to harmonize other information available across different countries in order to assess polycentricity in a more comprehen-sive way, also accounting for more functional dimensions. Second, compiling annual time series for the relevant variables on the FUAs would allow longitudinal analyses to be carried out, providing more robust empirical evidence on the relevance of spatial struc-tures in shaping socio-economic processes of countries, regions and metropolitan areas.

Assessing spatial structure and polycentricity has different potential implications for policy making at the three spatial levels considered. At the metropolitan level, the way people and economic activities are distributed across space raises important issues of effi-ciency in terms of public service provision, face-to-face interactions among economic agents, efficiency of transport and environmental issues connected with patterns of land development (e.g. sprawl). At the regional scale, spatial structures can influence economic development. More specifically, the presence of large metropolitan areas enhances agglomeration economies and consumption benefits. A polycentric spatial structure characterized by a network of cities has been thought to have the potential to compensate for the advantages of a single large agglomeration (borrowed size hypothesis, see also Burger *et al.*, 2014). However, our preliminary empirical exploration suggests that lower polycentricity and higher urban primacy are associated with higher GDP per capita. This may suggest that physical distance and agglomeration of people and workers have an important role for socio-economic conditions in regions. On the other hand, at the national level, more polycentric structures are found to be positively associated with GDP per capita. These first descriptive results suggest that the meaning and the policy significance of polycentricity are strongly affected by the scale at which it is observed. In economic terms, a balance between monocentricity at the regional level and polycentricity at wider spatial scales might be assumed. Polycentricity has been representing a key policy goal in the European policy discourse, where polycentric development is also seen as a policy tool to reach a more balanced development, hence lower territorial disparities. However, the main message emerging from this paper is that policy aimed at fostering polycentricity should be referred to specific spatial scales and should be more informed on the possible implications in economic and environmental respects.

Acknowledgements

The authors would like to thank Joaquim Oliveira Martins, Rudiger Ahrend, Karen Maguire, William Tompson and two anonymous referees for their helpful comments and discussions. Authors thank also, for their valuable comments, Roberto Camagni, Roberta Capello and all the participants in the International Seminar on "Welfare and Competitive-ness in the European polycentric urban structure: which role for metropolitan, medium and small cities?" that took place in Florence on June 2013. The views expressed in this paper are those of the authors and do not reflect those of the OECD or its member countries.

Notes

1. The Sprawl Index could not be computed in Canada, Chile, Korea and Mexico due to the absence of the land use layer in two points in time for these countries.
2. Among the countries included in the analysis, those considered as Eastern European ones are Czech Republic, Estonia, Hungary, Poland, Slovenia and Slovak Republic.
3. In any case, all models show a very low Variance Inflation Factors, which suggest no problems of multi-collinearity.
4. All the variables have been computed from the OECD database (http://dotstat.oecd.org/Index.aspx).

References

Alonso, W. (1964) *Location and Land Use* (Cambridge: Harvard University Press).

Alonso, W. (1973) Urban zero population growth, *Daedalus*, 102(4), pp. 191–206.

Anas, A., Arnott, R. & Small, K. A. (1998) Urban spatial structure, *Journal of Economic Literature*, 36(3), pp. 1426–1464.

Bailey, N. & Turok, I. (2001) Central Scotland as a polycentric urban region: Useful planning concept or chimera? *Urban Studies*, 38(4), pp. 697–715.

Brueckner, J. K. (2001) Urban sprawl: Lessons from urban economics, in: J. R. Pack & W. G. Gale (Eds) *Brookings Wharton Papers on Urban Affairs*, pp. 65–97 (Washington, DC: Brookings Institution Press).

Burger, M. & Meijers, E. (2012) Form follows function? Linking morphological and functional polycentricity, *Urban Studies*, 49(5), pp. 1127–1149.

Burger, M., Meijers, E., Hoogerbrugge, M. & Masip Tresserra, J. (2014) Borrowed size, agglomeration shadows and cultural amenities in North-West Europe, *European Planning Studies*. doi:10.1080/09654313.2014.905002

Cervero, R. (2001) Efficient urbanisation: Economic performance and the shape of the metropolis, *Urban Studies*, 38(1), pp. 1651–1671.

Champion, A. G. (2001) A changing demographic regime and evolving polycentric urban regions: Consequences for the size, composition and distribution of city populations, *Urban Studies*, 38(4), pp. 657–677.

Cheshire, P. (1999) Trends in sizes and structures of urban areas, in: P. C. Cheshire & E. S. Mills (Eds) *Handbook of regional and urban economics*, ch. 35, pp. 1339–1373 (Amsterdam: North Holland).

Combes, P., Duranton, G., Gobillon, L., Puga, D. & Roux, S. (2012) The productivity advantages of large cities: Distinguishing agglomeration from firm selection, *Econometrica*, 80(6), pp. 2543–2594.

Davoudi, S. (2003) Polycentricity in European spatial planning: From an analytical tool to a normative agenda, *European Planning Studies*, 11(8), pp. 979–999.

De Goei, B., Burger, M. J. & Van Oort, F. G. (2010) Functional polycentrism and urban network development in the greater south east, United Kingdom: Evidence from Commuting Patterns, 1981–2001, *Regional Studies*, 44(9), pp. 1149–1170.

Dieleman, F. M. & Faludi, A. (1998) Polynucleated metropolitan regions in Northwest Europe: Theme of the special issue, *European Planning Studies*, 6(4), pp. 365–377.

Dijkstra, L., Garcilazo, E. & McCann, P. (2013) The economic performance of European cities and city-regions: Myths and realities, *European Planning Studies*, 21(3), pp. 334–354.

Duranton, G. (2007) Urban evolutions: The fast, the slow, and the still, *American Economic Review*, 97(1), pp. 197–221.

Duranton, G. & Puga, D. (2004) Microfoundations of urban agglomeration economies, in: J. V. Henderson & J. F. Thisse (Eds) *Handbook of Regional and Urban Economics*, ed. 1, vol. 4, chap. 48, pp. 2063–2117 (Amsterdam: North Holland).

Eeckhout, J. (2004) Gibrat's law for (all) cities, *American Economic Review*, 94(5), pp. 1429–1451.

ESDP. (1999) Towards Balanced and Sustainable Development of the Territory of the EU. Presented at the Informal Meeting of Ministers Responsible for Spatial Planning of the Member States of the European Union, Postdam.

ESPON 3.1. (2003) From Project Results to "ESPON Results". A Draft Guidance paper Prepared by ESPON 3.1. Federal Office for Building and Regional Planning, Bonn.

Frey, H. (1999) *Designing the City. Towards a More Sustainable Form* (London: E&FN Spon).

Gabaix, X. (1999) Zipf's law for cities: An explanation, *Quarterly Journal of Economics*, 114(3), pp. 739–767.

103

Gabaix, X. & Ioannides, Y. M. (2004) The evolution of city size distributions, in: J. V. Henderson & J. F. Thisse (Eds) *Handbook of Regional and Urban Economics*, ch. 53, pp. 2341–2378 (Amsterdam: North Holland).

García López, M. A. & Muñiz, I. (2013) Urban spatial structure, agglomeration economies, and economic growth in Barcelona: An intra-metropolitan perspective, *Papers in Regional Science*, 92(3), pp. 515–534. doi:10.1111/j.1435-5957.2011.00409.x

Glaeser, E. L. & Kahn, M. E. (2004) Sprawl and urban growth, in: V. Henderson & J. F. Thisse (Eds) *Handbook of Regional and Urban Economics*, Vol. 4, Chap. 56, pp. 2481–2527 (Amsterdam: North Holland).

Glaeser, E. L., Kolko, J. & Saiz, A. (2001) Consumer city, *Journal of Economic Geography*, 1(1), pp. 27–50.

Green, N. (2007) Functional polycentricity: A formal definition in terms of social network analysis, *Urban Studies*, 44(11), pp. 2077–2103.

Hsu, W.-T. (2012) Central place theory and city size distribution, *The Economic Journal*, 122(563), pp. 903–932.

Kloosterman, R. C. & Musterd, S. (2001) The polycentric urban region: Towards a research agenda, *Urban Studies*, 38(4), pp. 623–633.

Lang, R. E. (2003) *Edgeless Cities: Exploring the Elusive Metropolis* (Washington, DC: Brookings Institution Press).

Meijers, E. & Burger, M. J. (2010) Spatial structure and productivity in US metropolitan areas, *Environment and planning A*, 42(6), pp. 1383–1402.

Meijers, E. & Sandberg, S. (2008) Reducing regional disparities by means of polycentric development: Panacea or Placebo, *Scienze Regionali*, 7(2), pp. 71–96.

Mills, E. S. (1972) *Studies in the Structure of the Urban Economy* (Baltimore, MD: John Hopkins Press).

Modarres, A. (2011) Polycentricity, commuting pattern, urban form: The case of Southern California, *International Journal of Urban and Regional Research*, 35(6), pp. 1193–1211.

Muth, R. (1969) *Cities and Housing* (Chicago: University of Chicago Press).

Nordregio. (2005) *Potentials for Polycentric Development in Europe*. ESPON Final Report 1.1.1, Luxembourg.

OECD. (2012a) *Redefining "Urban". A New Way to Measure Metropolitan Areas* (Paris: OECD Publishing).

OECD. (2012b) *Compact City Policies. A Comparative Assessment* (Paris: OECD Publishing).

OECD. (2013a) *Regions at a Glance* (Paris: OECD Publishing).

OECD. (2013b) *Rural-Urban Partnerships: An Integrated Approach to Economic Development* (Paris: OECD Publishing).

Parkinson, M., Meegan, R. & Karecha, J. (2014) City size and economic performance: Is bigger better, small more beautiful or middling marvellous? *European Planning Studies*. doi:10.1080/09654313.2014.904998

Parr, J. B. (2002) Agglomeration economies: Ambiguities and confusions, *Environment and Planning A*, 34(4), pp. 717–731.

Parr, J. B. (2004) The polycentric urban region: A closer inspection, *Regional Studies*, 38(3), pp. 231–240.

Priemus, H. (1994) Planning the randstad between economic growth and sustainability, *Urban Studies*, 31(3), pp. 509–534.

Riguelle, F., Thomas, I. & Verhetsel, A. (2007) Measuring urban polycentrism: A European case study and its implications, *Journal of Economic Geography*, 7(2), pp. 193–215.

Rosenthal, S. & Strange, W. C. (2004) Evidence on the nature and sources of agglomeration economies, in: J. V. Henderson & J. F. Thisse (Eds) *Handbook of Regional and Urban Economics*, Ed. 1, Vol. 4, Chapt. 49, pp. 2119–2171 (Amsterdam: North Holland).

Schwanen, T., Dieleman, F. M. & Dijst, M. (2004) The impact of metropolitan structure on commute behavior in the Netherlands: A multilevel approach, *Growth and Change*, 35(3), pp. 304–333.

Vandermotten, C., Roelandts, M. & Cornut, P. (2007) European polycentrism: Towards a more efficient and/or more equitable development? in: N. Cattan (Eds) *Cities and Networks in Europe. A Critical Approach of Polycentrism*, pp. 51–64 (Mountrouge: John Libbey Eurotext).

Veneri, P. (2010) Urban polycentricity and the costs of commuting: Evidence from Italian metropolitan areas, *Growth and Change*, 41(3), pp. 403–429.

Veneri, P. (2013a) The identification of sub-centres in two Italian metropolitan areas: A functional approach, *Cities*, 31, pp. 167–175.

Veneri, P. (2013b) On City-Size Distribution: Evidence from OECD Functional Urban Areas. *OECD Regional Development Working Papers* 2013/27, OECD Publishing.

Veneri, P. & Burgalassi, D. (2011) Spatial Structure and Productivity in Italian NUTS-3 Regions. Working Papers of the Department of Economics n. 364, Marche Polytechnic University, Ancona, Italy.

Veneri, P. & Burgalassi, D. (2012) Questioning polycentric development and its effects. Issues definition and measurement for the Italian NUTS-2 regions, *European Planning Studies*, 20(6), pp. 1017–1037.

Wegener, M. (2013) Polycentric Europe: More Efficient, more Equitable and more Sustainable? Paper presented at the International Seminar on Welfare and competitiveness in the European polycentric urban structure, Florence 7 June 2013.

White, M. J. (1999) Urban areas with decentralised employment: Theory and empirical work, in: E. S. Mills & P. Cheshire (Eds) *Handbook of Regional and Urban Economics*, pp. 1375–1412 (Amsterdam: Elsevier Science).

First- and Second-Tier Cities in Regional Agglomeration Models

CHIARA AGNOLETTI, CHIARA BOCCI, SABRINA IOMMI,
PATRIZIA LATTARULO & DONATELLA MARINARI

Istituto Regionale Programmazione Economica Toscana (IRPET), Firenze, Italy

ABSTRACT *This work has the purpose of inquiring into the presence of an urban hierarchy within second-tier city areas and alternative agglomeration models differing in their self-propelling ability and territorial sustainability. To this aim we confront regional polycentric areas, by going inside the traditional agglomeration and variety economies and the land settlement model of small–medium urban poles. In particular, the present work compares four Italian regions characterized by a territorial development driven by second-tier cities. The first two sections of the paper evaluate the functional pattern of the different urban systems and subsequently measure their rank in terms of extra-regional attractiveness on demand, which is expressed by rare services (Sections 2 and 3). Sections 4 and 5 tackle the issue of sustainability of settlements by taking into account land consumption and the degree of territorial fragmentation caused by different urbanization models. We discovered good urban performances and settlement sustainability of the second-tier cities agglomeration model in Italian regions, which is stronger when based on the co-presence of specialized small cities (which can assure a minimum amount of local demand for advanced services) and a multifunctional medium urban centre (which can ensure rarer functions). These findings bring strong recommendations on urban policies.*

1. Introduction

The fundamental idea underlying urban economics is that the economic performance of different urban poles and their regions is influenced by the number, size and functional composition of the cities as well as by their urban hierarchy. In short, economic analysis is concerned with the degree to which a territorial concentration of human activities exerts a multiplicative effect on single outcomes, thus generating advantages called agglomeration economies, or urbanization economies when referred to the urban environment, whose main feature is variety. Many studies of the city focused on a particular quality that improves the advantages of urban agglomeration, i.e. its demographic dimension. However, it seems reasonable to expect that the efficient size of urban poles varies according to the different contexts, being affected by the functions, dimensions, reciprocal distances and levels of interaction with other cities. This sort of consideration is common to the theory of city networks (cf. Dematteis, 1992; Camagni & Salone, 1993; Camagni & Capello, 2000; Capello 2002) and to the studies of polycentric systems and second-tier cities (cf. Kloosterman & Lambregts, 2001; Davoudi, 2003; Parr, 2004; Meijers, 2007, Parkinson *et al.* 2014). Lastly, it also seemed that the rationale behind a territorial organization of human activities (number and size of cities, mutual relationships, and so on) changes over time together with the levels of technological development, the ways of transport, the dominant productive systems and the size of competition markets (Camagni, 1993). Presently, with the development of the tertiary sector and the globalization of markets, a fairly widespread opinion is that the city can once again be considered a key economic subject; its competitiveness relies on factors associated with the *scale* and the *productive mix* (hence, the supply of a wide range of public utilities and advanced services to firms), as well as on an adequate stock of transport and communication *infrastructures* and a more general *urban quality*, where the latter involves high architectural standards in public and private spaces and valuable educational, cultural and recreational services—which attract qualified human resources (cf. Florida, 2003; Burger *et al.*, 2014).

Moreover, in recent years urban development has often been accompanied by higher costs of sprawl and spoil of territorial resources. Increasing returns of scale, agglomeration and urbanization economies can, in fact, generate high land consumption and waste of territorial resources. In addition, urbanized land has grown independently from economic and population growth.

The aim of this work is to inquire into the performances of second-tier cities, looking for different hierarchical models within polycentric areas. Main attention is paid to the existence of urban hierarchies in regions driven by second-tier cities, and their effects on competitiveness and sustainability.

Urban and regional performances are analysed under the specific composition of urban functions and their attractiveness on extra-regional demand as a sign of the urban ranking, and compared with the different land settlements.

In particular, the analysis compares four Italian regions driven by second-tier cities, taking into consideration three issues:

(1) urban ranking, which is measured through the presence and distribution of rare urban functions (Section 2);
(2) urban competitiveness, which is the capacity to attract extra-regional demand of specialized functions, and the "urban hierarchy" effect deriving from the ability to export goods (Section 3);
(3) sustainability of settlements, by taking into account land consumption and the degree of territorial fragmentation connected to the different urbanization models (Sections 4 and 5).

The final result, illustrated in the conclusion (Section 6), is intended to point out how conditions in second-tier cities' patterns are consistent with the traditional agglomeration and diversification economies (competitiveness) and with a sustainable use of scarce natural resources (sustainability).

2. Urban Ranking: The City and Its Rare Functions

The first step for an analysis aimed at interpreting the urban structure is to identify the unit of analysis, in other words to define what we mean by city. As known, industrial development, and the related experiences of urbanization first and of suburbanization later, made the notion of a city's administrative boundaries outdated. In fact, these have been usually traced at a time when the compactness of settlements, the density of population and the concentration of economic activities went hand-in-hand, and were immediately visible as opposed to rural areas, characterized by low-intensity settlements. The modern city has a somewhat diluted shape instead, which poses problems of identification.

The procedure suggested in the present work is meant to circumscribe *city's boundaries* starting from a conceptual clarification of the unavoidable "ingredients" of an urban pole: *demographic dimension, economic dimension, quality and variety of functions* performed and a certain degree of *settlement compactness*. Consequently, the factual data on agglomeration modalities are not entirely neglected, but—as shown afterwards—they are employed at a later stage for comparison with other parameters. Our method has many points in common with those of several recent studies of urban structure. We can mention, first of all, the analyses carried out in the framework of the European project ESPON (Nordregio, 2004), then the later contributions inspired by them, and finally, for the Italian context, the surveys by Istituto Nazionale di Statistica (ISTAT) (2006), Bank of Italy (Di Giacinto *et al.*, 2012) and some university research teams (Calafati, 2012). The methodological innovation firstly identifies and classifies the cities and secondly ranks the urban hierarchy, through the repeated filtering of a set of data drawn from the mapping of the local labour systems (LLSs).[1]

The application of a demographic threshold introduces a first separation between urban and non-urban systems, where the former are by definition characterized by a stronger concentration of population. For the sake of methodological homogeneity, we choose to apply the threshold already used by ESPON (2007) contributions at the European level for the identification of *functional urban areas* (FUAs): these are local systems with a total population of no less than 50,000 inhabitants and a main centre of at least 15,000 inhabitants. This threshold is clearly very low, so that it leaves the ground for the application of the following filters.

The next step is *identifying urban functions*, a criterion that will be employed later to further select the territorial areas. Even this aspect is not new from a conceptual viewpoint, although a well-established method for identifying and selecting urban functions does not really exist in the Italian context of analysis at present. In general, the starting point is the commonly shared and empirically founded idea that cities, just because of their demographic and economic dimensions, can put into effect some unique and highly specialized functions, which cannot be found everywhere, but only in a few poles benefitting from very peculiar contextual conditions (i.e. urbanization economies). Highly specialized functions, which are identified on the basis of the employees belonging to the different economic sectors,[2] can then (i) provide a distinction between urban and non-urban functions and (ii) be assigned a different degree of concentration among urban areas. Among the urban functions selected, we find a number of *high- and medium-tech manufacturing activities*, as well as *highly specialized tertiary activities* associated with logistics, financial intermediation, academic education and research, health and business services. The activities associated with *academic education, telecommunication* and *publishing* seem particularly rare even in the urban environment (Table 1).

Once the urban functions are identified and thus arranged consistently with the wide literature on the Jacob's approach,[3] we introduced an additional component in the concept of city, the *variety* of the functions performed (cf. Van Oort *et al.*, 2014). FUAs are accordingly classified and arranged as to their *degree of urban specialization*, as well as to the *number of urban functions* in which they gain a specialization, the *quality* of these urban functions (rarity level), and the *homogeneity* of their specialization

Table 1. Classification of urban functions (2009)

	Specialization index FUA/non-FUA (a)	Concentration index normalized for urban FUAs (b)	Synthetic index of urban functions (geometric average of a*b)
HT and ICT manufacturing			
Chemical-pharmaceutical industry	2.54	1.42	1.90
Computers, electronics, optics	1.91	1.23	1.53
MHT manufacturing			
Oil products	2.11	1.33	1.68
Chemical products	1.70	1.15	1.40
Mechanical engineering	1.14	1.10	1.12
Means of transport	1.29	1.16	1.22
Logistics			
Transportation and storage	1.39	1.01	1.18
Financial services			
Financial and insurance activities	1.72	1.03	1.33
Publishing and telecommunications			
Publishing	5.18	1.31	2.61
Telecommunications	8.05	1.33	3.27
KIBS and universities			
Information services	2.45	1.16	1.68
Professional activities	1.32	0.97	1.13
Research & Development	2.62	1.23	1.80
Other professional activities	1.58	1.04	1.29
University staff	11.10	1.37	3.90
Personal care services			
Real estate activities	1.11	0.89	0.99
Arts and entertainment activities	1.05	0.98	1.01
Public utilities (water, urban sanitation)	1.71	0.98	1.30
Health Services	1.34	1.03	1.17
Other specialized services to firms			
Administration and support activities	2.27	1.07	1.56

Source: IRPET elaboration on data from ISTAT, Ministry of Education, Universities and Research and Ministry of Health.

Note: HT, high technology; ICT, information and communication technology.

level. Two simple indicators of economic performance (employees per inhabitants and value added per inhabitants), have finally been included in order to obtain a classification of urban poles which simultaneously takes into account the functions performed and the results obtained (Table 2).

The last filter is given by the data on the *agglomeration modalities*. We referred to an indicator which is available at European level, the *urban morphological zone* (UMZ), which is constituted by urbanized areas with a distance of no more than 200 m, and it was developed starting from the *Corine Land Cover* 2006 survey. We thus combined FUAs interconnected by a compact built fabric, in the assumption that, even if it is true that the contemporary city no longer corresponds to a compacted built area but can be better approximated as the territory in which socioeconomic relationships are more dense (which is the reason why we initially decided to adopt a functional definition), it is also true that these relationships are present and significant where they take the form of a morphological category. The presence of UMZs cross-cutting the boundaries of FUAs can, however, be observed in Italy only in a few cases, and specifically for our research areas in: Thiene and Schio (197,000 inhab.) in Veneto, the Rimini coast (298,000 inhab.) in Emilia-Romagna, Florence and Prato (985,000 inhab.) and the Versilia coast (523,000 inhab.) in Tuscany.[4]

By summarizing all the relevant aspects above, we obtain a hierarchical classification of the urban poles existing in the national territory (Table 3), which allows us to compare the agglomeration structures of the four regions with a prevalence of second-tier cities which are under exam: Veneto, Emilia-Romagna, Tuscany and Marche.

The indicators were summarized using the first factor of a principal component analysis, whose aim is inferring a latent variable (urban ranking) from numerous observable variables. The afore-mentioned variance, the correlations between the initial variables and the first component, as well as the contribution of each variable to the final result are illustrated in Table 4. The extracted factor, assumed as the index of urban ranking, explains a 52% of the overall variance and derives from basically equal contributions of the initial variables.

Consistently with their agglomeration structure, mainly based on small- to medium-sized cities, the considered regions have no large metropolitan system (over one million inhabitants), which instead can typically be found in regions with a monocentric urban system (Lombardy, Piedmont, Lazio and Campania); a characteristic that does not

Table 2. Indicators of urban hierarchy in FUAs

Criteria for FUA ranking	Indicators
Specialization in urban functions	Employees in rare urban functions on total amount compared to FUAs average
Quality of urban functions	Rarity level of existing functions (each function has a different weight)
Functional variety	No. of existing rare functions
Sectorial dispersion/concentration of urban specialization	Inverse of the coefficient of variation (CV) for sector specialization
Economic performance	Employees per inhabitant and added value per inhabitants

Table 3. The urban hierarchy in Italy

City typology	Population size class	City—LLS or total of LLSs (descending order of population)	Presence of urban functions (high > 3°quartile, low < 1°quartile)	Economic performance (high > 3°quartile, low <1°quartile)	Specialization index of production functions	Specialization index of cultural functions	Synthetic urban ranking index (urban if ≥0.5)
Large metropolitan systems	>1,000,000	Milan area[a]	High	High	1.2	1.3	2.95
		Rome	High	High	1.0	1.7	2.78
		Turin	High	High	1.4	1.3	2.37
		Naples area[a]	High	Low	0.9	1.0	0.78
Middle-sized metropolitan systems	500,000–1,000,000	Bologna	High	High	1.3	1.2	3.34
		Genoa	High	High	0.9	1.3	2.41
		Florence—Prato area[a]	High	High	0.8	1.0	2.10
		Padua	High	High	1.0	1.2	1.86
		Venice	High	High	0.9	0.8	1.51
		Verona	High	High	0.8	0.9	1.46
		Bari	High	Medium	1.2	1.2	1.41
		Brescia-Lumezzane Area[a]	Medium	High	1.2	0.9	1.39
		Catania–Acireale Area[a]	High	Medium	1.0	1.0	1.01
		Bergamo–Albino Area[a]	Medium	High	1.2	0.7	0.97
		Palermo–Bagheria Area[a]	High	Medium	0.8	1.2	0.61
		Parma	High	High	1.3	1.0	1.74
		Modena	High	High	1.4	0.8	1.40

(Continued)

Table 3. Continued

City typology	Population size class	City—LLS or total of LLSs (descending order of population)	Presence of urban functions (high > 3° quartile, low < 1° quartile)	Economic performance (high > 3° quartile, low < 1° quartile)	Specialization index of production functions	Specialization index of cultural functions	Synthetic urban ranking index (urban index if ≥ 0.5)
Medium cities	250,000–500,000	*Reggio Emilia*	High	High	1.5	0.8	1.36
		Vicenza	High	High	1.2	0.8	1.26
		Udine	High	High	1.1	1.0	1.22
		Cagliari	High	Medium	1.0	1.2	1.13
		Pescara	Medium	Medium	1.0	1.0	0.81
		Lecce	High	Medium	0.8	1.3	0.69
		Treviso	Medium	High	0.8	0.8	0.64
		Salerno	High	Medium	0.7	1.1	0.55
		Frosinone	Medium	Medium	1.1	0.6	0.55
		Pisa	High	High	0.9	1.9	1.91
		Siena	High	High	0.8	1.5	1.86
Small cities[b]	100,000–250,000	*Ancona*	High	Medium	1.0	1.1	1.64
		Ferrara	High	High	1.0	1.1	1.33
		Livorno	High	Medium	1.1	0.7	0.96
		Pesaro	Medium	High	1.1	1.0	0.90
		Piacenza	High	High	1.2	0.8	0.85
		Forlì	Medium	High	1.0	0.7	0.64
		Ravenna	Medium	High	0.8	0.7	0.72
		Sassuolo	Medium	High	1.3	0.4	0.66
		Lucca	Medium	High	0.7	0.8	0.63
		Cesena	Medium	High	0.8	0.7	0.56

Source: IRPET elaboration on data from ISTAT, Ministry of Education, Universities and Research and Ministry of Health.

Note: Italics show the cities belonging to regions based on second-tier cities patterns.

[a]These areas combine two or more urban LLSs, united in a same UMZ. In details, the Milan area is the sum of the LLSs of Milan, Busto Arsizio, Como, Seregno, Sesto Calende and Varese, while the Naples area comprises the LLSs of Naples, Aversa, Castellammare di Stabia, Cava de' Tirreni, Nocera Inferiore, Nola and Torre del Greco.

[b]The scored cases are only those of the four regions under analysis. There are still 19 cases in the remaining regions; only 11 cases among the class of population under 100,000 inhabitants have an urban ranking, five of which are in polycentric regions.

Table 4. Results of the principal component analysis

Components	Eigenvalues	% Variance	Cumulated %	Variables	Coefficients of correlation with the first component	Factorial weights for the first component
1	3.098	51.6	51.6	Abs. value per inhabitant	0.821	0.265
2	1.250	20.8	72.5	No. of urban specializations	0.782	0.252
3	0.975	16.3	88.7	Total urban specializations	0.774	0.250
4	0.393	6.6	95.3	Employees per inhabitant	0.710	0.229
5	0.208	3.5	98.7	Mark for urban functions	0.676	0.218
6	0.076	1.3	100.0	1/CV urban specializations	0.505	0.163

Source: IRPET elaboration on data from ISTAT, Ministry of Education, Universities and Research and Ministry of Health.
Note: % Variance explained by the first component: 51.6.

always correspond to a high level of development, as the case of the Naples metropolitan area well illustrates. On the other hand, the cities belonging to less agglomerated regions gain a steady position at the intermediate levels of the urban hierarchy, that is among the medium-sized metropolitan systems (5 cases out of 11) and the medium-sized cities (5 cases out of 11); some excellent cases can also be observed among the small cities (Pisa, Siena and Ancona, which together with Trento, Bolzano and Trieste combine a small dimension with an urban index ranking above 1.5).

On the basis of the characteristics highlighted for each pole it is possible to build the *regional profiles* (Table 5).

Among the cases under study, *Veneto* seems the region lacking a leading centre the most, since it presents three medium-sized metropolitan systems (Padua, Venice and Verona) instead of one, and whose demographic dimension is much similar and urban ranking rather good. Still, the latter is not so high, and eventually the city reaching a highest rank (Padua) is not the regional capital. The urban hierarchy is completed by two medium cities with a low urban ranking, being specialized in traditional manufactur-

Table 5. Illustration of urban hierarchy in regions with second-tier cities

	Synthetic urban ranking index																			
	Veneto					Emilia-Romagna					TUSCANY					Marche				
Population	0.5-1	1-1.5	1.5-2	2-3	>3	0.5-1	1-1.5	1.5-2	2-3	>3	0.5-1	1-1.5	1.5-2	2-3	>3	0.5-1	1-1.5	1.5-2	2-3	>3
> 1,000,000																				
500,000-1,000,000			XXX							X				X						
250,000-500,000	X	X						XX	X											
100,000-250,000						XXXXX							XX			XX		X		X	

Source: IRPET elaboration on data from ISTAT, Ministry of Education, Universities and Research and Ministry of Health.
Note: Cities (x) classified for urban ranking (rows) and population (columns). The cells with a thin frame stand for high urban rank, while those with a thick frame stand for very high urban rank.

ing activities. Therefore, Veneto has many rich and highly populated areas that look more like districts rather than cities.

The urban hierarchy of *Emilia-Romagna* can be considered the most comprehensive. It has one medium-sized metropolitan system with a very high urban ranking (Bologna), three medium cities with a parallel demographic weight and a good quality of urban functions (especially in the case of Parma), and five small towns with a low urban ranking, having functions more on the manufacturing side.

The metropolitan system of *Tuscany* is of medium extent, very densely populated and of a medium–high urban ranking; it has two constituents (Florence and Prato) with very different functional specializations—urban functions for Florence and traditional manufacturing for Prato. The evidenced features confirm that these two cities combined in a single pole because of the short geographical distance, and not through a full integration deriving from a process of urban polarization. It also seems worth pointing out that, even considering the urban pole of Florence's FUA alone, its population exceeds the 700,000 inhabitants and its urban ranking is higher than 2.5. This makes us infer that in any case the Tuscan system presents an urban apex with dimensions and functional characteristics that are comparable to those of the Bologna area in Emilia-Romagna. The central pole of Tuscany is relatively lacking in specialized services for manufacturing, as regards its urban functions. The regional urban hierarchy lacks medium-sized cities, while there are small cities hosting excellent functions, Pisa and Siena. The first town, due in particular to a good endowment of transport infrastructures, constitutes a "positive anomaly", being an important doorway to the region (cf. Cirilli & Veneri, 2012).

Finally, the less densely populated region among those examined here, *Marche*, has no urban pole at the top of the hierarchy, but only two small cities (Ancona and Pesaro) of which only the first presents a significant urban ranking. Hence, this region is predominantly composed of small manufacturing districts.

3. Urban Competitiveness: The Ability to Export Rare Functions

Instead of starting a deeper analysis of other relevant features in the definition of regions with small- to medium-sized cities—such as the existence of stable and constant relationships among poles—we draw attention to the assessment of regional competitiveness in terms of urban functions. Consequently, we intend to match the features of the agglomeration structure in the four regions analysed against the export performances achieved for three high-ranking urban functions: highly specialized health services (hospital services), specialized education opportunities (academic education) and, as regards the productive world, training opportunities or work meetings and the like. To measure the degree of attractiveness for the different regional urban systems, we use in the first case the quantity of admissions of patients from outside the region, in the second case the number of the local university's enrolments of students from outside the region, and in the third case the amount of business trips directed to the region.

Finally, we test the effect of urban hierarchy on the ability to export goods.

3.1. *Attractiveness of Health Services*

In Italy, patient mobility among regions moves more than 800,000 people every year for a total of 6.5 millions hospital admissions. Considering that the administration of health care is assigned to regional governments, that flows have a marked south-to-north directional-ity, and that the average cost for hospital admission is higher in the most attractive regions, this figure can very well be considered as an excellent indicator of the quality of the ser-vices supplied. In other words, the demand for health assistance outside the region can be viewed as a search for higher specialization in health care.

Using a synthetic index of attractiveness such as the ratio of incoming to outcoming hos-pitalization flows, we find that the best result is obtained by Emilia-Romagna, that per-forms this function more efficiently than the second-best region, Lombardy, which is more densely populated, has a comparable level of development and a monocentric struc-ture. At the third position is another medium-sized urban structure, Tuscany. In the national classification, Veneto and Marche come in the eighth and ninth places, respect-ively (Table 6).[5]

The ability to attract patients from other regions probably derives from a multitude of factors, such as the overall level of supply (the number of beds per inhabitant), its organ-ization in not-too-small poles (with over 20,000 inhab.), the ease of access through the major transport routes or the general quality of the services provided. The disparity between Emilia-Romagna and Tuscany may be explained with the smaller number of medium poles existing in the second region (where, probably not by chance, the supply is concentrated in the two great poles of Florence and Pisa), while Veneto probably

Table 6. Attractives of urban functions in the Italian regions

Hospital function 2011 (a)		University function 2009 (b)		Business trips 2004–2008 (c)	
Attractiveness index		Attractiveness index		Attractiveness index	
Emilia-Romagna	2.59	Emilia-Romagna	5.27	Emilia-Romagna	1.65
Lombardy	2.17	Tuscany	4.74	Lazio	1.54
Tuscany	1.79	Lazio	3.35	Umbria	1.47
Friuli V.G.	1.56	Lombardy	3.21	Lombardy	1.42
Lazio	1.35	Umbria	2.15	Trentino-Alto Adige	1.34
Molise	1.33	Friuli-Venezia Giulia	2.03	Abruzzo	0.94
Umbria	1.23	Abruzzo	1.94	Veneto	0.94
Veneto	1.16	Marche	1.44	Marche	0.89
Marche	0.94	Piedmont	1.20	Tuscany	0.87
Trentino-Alto Adige	0.93	Liguria	0.85	Liguria	0.84
Liguria	0.91	Puglia	0.82	Piedmont	0.79
Piedmont	0.86	Trentino-Alto Adige	0.78	Molise	0.79
Basilicata	0.68	Veneto	0.77	Basilicata	0.73
Abruzzo	0.66	Molise	0.45	Friuli-Venezia Giulia	0.72
Puglia	0.45	Sicilia	0.34	Sicilia	0.65
Valle d'Aosta	0.44	Valle d'Aosta	0.12	Sardinia	0.56
Sardinia	0.31	Basilicata	0.09	Valle d'Aosta	0.54
Campania	0.30	Sardinia	0.06	Campania	0.54
Sicilia	0.30	Calabria	0.06	Calabria	0.53
Calabria	0.13	Campania	0.05	Puglia	0.42

Source: IRPET elaboration on data from Ministry of Health (a), Ministry of Education, Universities and Research (b) and ISTAT, *I viaggi degli italiani.*
Note: Italics show the regions marked by the presence of second-tier cities.

suffers for its greater peripherality and lower urban ranking, and Marche has a position in line with the reduced dimension and ranking of its cities.

3.2. *Attractiveness of the University Function*

Being a high-level specialized activity, academic education is typically an urban function, usually located in medium to large cities rich in cultural and economic heritages.

The students enrolled in university courses outside their region are a little more than 480,000 of a total of 1,800,000. Among the most attractive regions in terms of students from other regions we find those that provide a larger supply (Lazio and Lombardy, with a total of 250,000 university students), but also a medium supply (like Emilia-Romagna, with a total of 150,000 enrolled students).

Again, size alone does not explain the differences in performance: in fact, regions where the supply of academic education is very significant, like Campania and Sicily, still present very moderate levels of attractiveness for students from outside regions.

If the index of attractiveness is instead calculated as the ratio of immigrants to emigrants, the two best positions in the ranking are occupied by two second-tier cities' regions with an all-comprehensive urban hierarchy, Emilia-Romagna and Tuscany (Table 6).

3.3. *Attractiveness of Business Trips*

Because cities are characterized by a high concentration of job opportunities, they are usually the centre of a significant portion of both daily work commuting and business trips. The difference between the two kinds of travels is that the second involves at least one overnight stay and is registered through sample data collected by ISTAT.[6] According to the figures surveyed for the period 2004–2008, the total business trips amounts to about 11 millions a year. The two regions attracting the majority of travels are those where the largest cities are located: Lombardy with Milan, and Lazio with Rome (17% both). At lower but still significant levels of attractiveness we find regions of Centre-North Italy, while the South clearly suffers the weakness of its production structure and still remains an origin rather than a destination of journeys. Emilia-Romagna achieves an excellent result, while the position of the residual regions in this ranking is affected by a productive fabric strongly centred on manufacture (Table 6). The only exception to this rule is Lazio, being the seat of the main national administrative functions.

3.4. *Ability to Export Goods*

As known, Italian exports of goods have a strong territorial concentration, as they mainly originate from central-northern regions, that is the areas where the presence of manufacturing is stronger and more stable over time. The highest values in terms of exports per inhabitant are achieved by Emilia-Romagna, Lombardy and Veneto, with an amount of about 9500 Euros per capita, followed by Piedmont (about 8000 Euros per inhabitant) and, at a greater distance, Tuscany and Marche (about 6000 Euros per inhabitant) (data from ISTAT). The ability to export goods is obviously highly correlated with the existence at the local level of manufacturing firms; of course, it is also true that the organization and degree of specialization of cities can add to competitiveness as far as they provide

Table 7. Export performance of regions with second-tier cities

	Coefficients	Significance value
Export value per inhabitant		
Specialized manufacturing	0.901	0.000 (***)
Urban ranking	0.312	0.010 (***)
Corrected R^2	67.1%	
Number of observations	70	

Source: IRPET elaborations on ISTAT data.
Note: Results of the regression analysis with standardized values.

upstream and downstream services that have a bearing on the quality of the goods produced, and consequently on their exportability (product and process innovation, trading and technical assistance, and so on). A regression analysis carried out for the Italian export regions (basically, the central-northern regions) confirms that the ability to export, measured in terms of the export value per inhabitant, is indeed influenced by the rank of the urban systems. As regards in particular the regions under analysis, the results are illustrated in Table 7.

About 67% of exports of manufacturing goods can be explained by the strong presence of manufacturing activities and the degree of urban ranking. The major contribution is obviously provided by the first explanatory variable, although the second variable's contribution is quite relevant and positive. Therefore, as the urban ranking grows, there is also an increase in the competitiveness of local systems on the international markets of goods.

4. The Sustainability of Agglomeration Patterns: Different Levels of Land Consumption

After examining the issues related to the urban structure, the range of its functions and its competitiveness, we investigate its impact in terms of land use and urbanization of agricultural areas. In fact, these aspects are assuming a growing importance in the assessment of territorial sustainability of agglomeration patterns.

4.1. Land Consumption in Regions with Second-Tier Cities

In order to better describe the types of regional agglomeration, we compare the different extents and distribution characteristics of urbanized areas.[7]

If we take into account the weight of urbanized areas over the total surface, the differences among regions seem rather marked: against an incidence for the whole country of 4.9%, the most land-consuming region is Veneto (8.2%), followed at a certain distance by Emilia-Romagna (4.9%), Tuscany and Marche, which achieve the same result (4.4%). If, in order to refine the investigation, we consider only the areas that are effectively attractive for building (thus leaving out mountain areas),[8] the gap between the two groups of regions is almost the same as before. The highest saturation level of land is registered in Veneto, while Tuscany's values are again lower than the national average and in line with Marche and Emilia-Romagna (Figure 1).

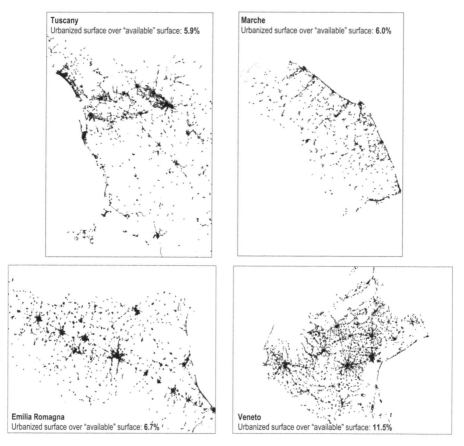

Figure 1. The regions' urbanized areas with a prevalence of second-tier cities (2006).
Source: IRPET elaboration on data from Corine Land Cover.

4.2. *Large Land Consumption in Last Decades*

Over the course of time, urbanization processes have experienced different dynamics and outcomes depending on their determinant factors. The main drivers can certainly be constituted by the demographic dynamics, broadly interpreted as the simple increase/decrease of population and the outcome of transformations in the family pattern, and the economic dynamics, which are positively correlated with urbanization. While in the past demographic dynamics might be used to describe the different stages of development of urban systems, with the emergence of post-Fordism the relationship that had traditionally existed between demographic and agglomeration processes loosened, and cities started to expand even where the growth of resident populations slowed to a standstill or even reverted. The agglomeration patterns resulting from the break of this relation are somehow unprecedented, so they cannot be described using the categories traditionally used in the urbanization literature.

There are also examples of settlements which grew along previously traced paths, which strengthened the conurbation processes already in place and thus caused the gradual "welding" of more places into a whole.

Among the main consequences of the development of these agglomeration modalities we find increase in land consumption, further expansion of private mobility and diffusion of an energy-consuming settlement pattern. Therefore, the widespread opposition to these new urban configurations is probably due to the higher economic and environmental costs associated with their development.

The afore-mentioned situation seems, however, particularly important during the 1990s, when the trend of population was no longer aligned with the increase in agglomerations. As a matter of fact, the two trends do not match in Tuscany, and in the other regions the growth of urbanized areas is always greater than the increase in population, except from Marche where inhabitants increased in number at a higher rate compared to urbanized areas. At the time, the per capita demand for urbanized land had a consistent growth, a fact that may be explained by a variety of reasons. Among them, we can certainly mention the transformation of the family pattern or, most importantly, the increase in the level of welfare, which results in a growing demand for second and even third holiday houses and in a change of lifestyles (apartment versus small house). Another explanation may be found in the advent, in that period, of a new generation of territorial planning tools, which contributed to widening the gap between demographic and agglomeration dynamics. In fact, the town-planning culture permeating these new plans has a propensity to identify the development of territory with growth *tout court*, and consequently with the increase in settlements. This results in expectations of growth for urban functions being highly overestimated and far from the actual trends experienced by territories.[9] In addition, the plans in force at that stage had been in the making for too long a time, which resulted almost everywhere—though with marked differences from case to case—in an exacerbated pressure on real estate markets, and an ensuing magnification of their performances. Furthermore, during the 2000s a new phase started, which was characterized by a moderate growth of agglomerations and different relation between demographic and settlement variables: at the national level, and particularly in Veneto, the average annual demographic growth is slightly above settlement development, while in Tuscany and Marche the opposite is valid. This is rather due to the increase in population than to restrictions on new urbanization (Figure 2).

The limitation of agglomeration expansion – where they occurred – may be ascribed to a variety of concurrent causes: on the one hand, the slowdown in the economic

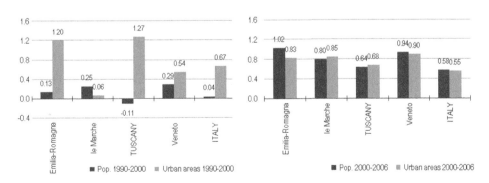

Figure 2. Average annual increases in urbanized surfaces and population in regions with a prevalence of second-tier cities (1990–2000 and 2000–2006, percentage values).
Source: IRPET elaboration on data from Corine Land Cover and ISTAT.

dynamics entailing a natural tendency to control settlement expansion, and on the other hand the emergence of a cultural attitude for urban recovery and requalification that has partly directed land-use administrations towards choices of conversion of the existing urban fabric. At the same time, we witness the appearance in regional regulations of land-use reforms based on the principle of sustainability, thus aimed at the containment of settlement expansion, even though their role in the reduction of land consumption has been somewhat ambiguous so far. The suggested shift in perspective, from a view of the territory as a mere support to development to considering it a resource, should have involved a radical change, theoretical and practical at the same time, in the modalities and extent of the forecasts put forward in territorial planning tools. In contrast, as evidenced by much research on the issue, such a shift has taken place only partially, and the methodological and theoretical references laid at the basis of growth assumptions turned out to be generally weak (IRPET, 2011, 2012). Consequently, all the forecasts, although more controlled compared to those made in the previous generation of plans, are still too wide and not always matched by the ongoing socioeconomic processes.

4.3. *Urbanization of Agricultural Land*

Among the crucial points raised by the literature as regards the growth of urbanized areas is the evidence that urbanization processes usually take place to the detriment of agricultural land use.

In Italy, the result of extensive urbanization during the period 2000–2006 is that 94% of the urbanized areas has taken the land away from agriculture (for a total of 46,000 ha). If we consider the ratio between the land converted from agricultural use and the population in the regions under analysis, we find that in Tuscany the removal of land from agriculture was the least extensive, while Veneto again registers the highest values. Marche and Emilia-Romagna are at an intermediate position, recording similar values. However, it is worth noting that all the investigated regions have values that are above the national average, a fact causing concern not only in terms of a possible loss of landscape quality resulting from cities' expansion, but also with respect to the environment protection and identity-making role traditionally held by agriculture. At a juncture in which the resources for land management are scarce, it is obvious that dropping the role of agricultural activities in securing land use can turn into a serious danger for our territorial systems (Figure 3).

We do not discuss here the quantification of the loss of agricultural land, but on the other hand we focus on the identification of typologies for developing urbanized areas. It seems useful to recall the significance of the introduction in our country, during the 1900s, of the new sale locations (malls) and entertainment venues. Even in the period 2000–2006, just like in the previous decade, the development of urbanized areas had mostly involved industrial and commercial places: in Italy, 60% of the new urban expansion took place in areas dedicated to commercial/industrial use. The disparities at the regional scale pointed out that the increase in the industrial typology was more significant in Marche (66%) and less in Tuscany (57%) (Figure 4). There is evidence for another crucial point together with these data, which is associated with the different territorial fabric to which settlement growth applies. More recently, urban development—as already mentioned in this work and underlined in the scientific literature—despite minor dissimilarities at the regional level, seems directed at enhancing a sort of urban fabric distinguished by

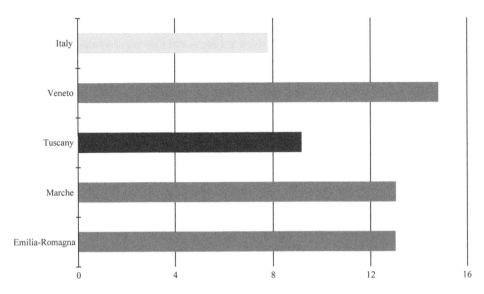

Figure 3. Agricultural land turned to urbanized areas in regions with a prevalence of second-tier cities (2000–2006, per capita sq km * 100).
Source: IRPET elaboration on data from Corine Land Cover and ISTAT.

little compactness, a more rarefied and discontinuous texture of settlements. This seems to prove how in the period analysed the periurban trends in housing choices are still active. This phenomenon can be mainly traced back to the so-called rental effect that has been particularly marked during the first half of the 2000s, when the slowdown of growth already upon Tuscany pushed up the demand for real estate investments. This clearly affected the cost of housing, and encouraged the relocation of some sections of the population to more peripheral areas, in search of a better cost-quality ratio in their housing standards.

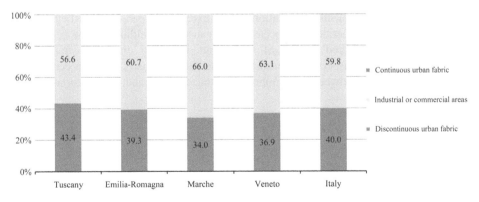

Figure 4. Classification of the new urban fabrics in regions with a prevalence of second-tier cities (2000–2006, settlement growth = 100).
Source: IRPET elaboration on data from Corine Land Cover and ISTAT.

5. Settlement Fragmentation and Sustainability

After having analysed the extension of urbanized areas and the recent dynamics of urbanization processes, we now look into the more specific morphological features of settlements, with particular regard to their fragmentation level (cf. Brezzi & Veneri, 2014). As now widely acknowledged, the sprawl causes a huge loss of land, given that the surface actually occupied by buildings represents only a share of the total settlement area; accordingly, another variable will determine the possibilities of land use: the level of fragmentation (Romano, 2004; Irwin & Bockstael, 2007; Romano *et al.*, 2010). As a matter of fact, fragmentation is today acknowledged as one of the main crucial factors of land urbanization. The incessant disintegration of settlements does indeed pose a two-fold problem: on one side, the reduction and/or disappearance of natural environments that affect the integrity of ecosystems, and on the other side the progressive "insularization" of residual spaces. Besides, a fragmented settlement pattern is not efficient because it wastes the land, raises the costs for the provision of services and increases private mobility, not being capable of promoting an economically sustainable management of public transport.

To measure the level of fragmentation engendered by settlements, we examined the areas previously defined as urban[10] and calculated the urban fragmentation index, which represents the density of an urbanized surface over the whole municipality area, weighted with a form factor (Romano & Paolinelli, 2007). In other terms, we focused on the surface occupied by settlements and, through the use of a geographic information system procedure, defined its perimeter and then compared it to the theoretical perimeter of the nearest most compact shape, i.e. a circle. The resulting indicator takes higher values where fragmentation is strong, and inversely lower values—approaching 1—where the shape of the urbanized zone is made of very few or just one aggregation, that tends to look like a circle.

The equation used to calculate the index is as follows:

$$\text{UFI} = \frac{\sum \text{Urbanized surface}}{\text{Territorial surface}} \times \frac{\sum \text{Urbanized perimeter}}{2\sqrt{\pi \sum \text{Urbanized surface}}}.$$

Although the process of suburbanization and the extensive expansion of settlements that started in the post-war period have occasionally caused a "welding" of the centre with the outskirts, the scattered feature of settlements is still a distinctive trait of our country. The index of dispersion, measuring the territorial density of urban centres,[11] is calculated as the ratio between the number of urbanized cores and the territorial surface:

$$\text{Dispersion} = \frac{\text{Number of urbanized cores}}{\text{Surface}}.$$

The two indicators calculated on a municipality basis reveal that fragmentation and dispersion describe the whole territory of Lombardy and Veneto, a large portion of Marche, and only the more urbanized areas of Emilia-Romagna and Tuscany. In these regions, we find that problematic areas are those around the main urban zones, since the indicators take higher values near the most important metropolitan systems, and particularly at the poles

(a)   (b)

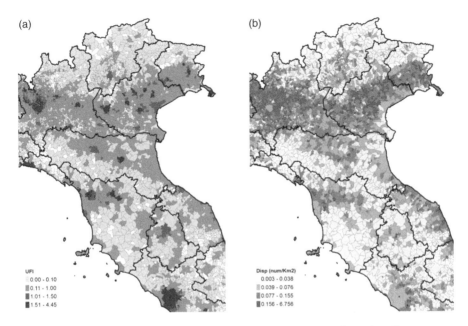

Figure 5. Urban fragmentation index (a) and settlement dispersion (b).
Source: IRPET elaboration on data from Corine Land Cover.

of these systems—where the saturation level raises—and at their ramifications along the plains and foothills, taking sparser values where the dispersion level is higher (Figure 5).

In this case again, we analyse settlement dispersion by calculating the number of urban centres for each municipality. This index gives back a mapping that does not differ much from the one obtained with the fragmentation level, apart from a better evidence for the extremely diffused feature of the Veneto settlement system.

On the contrary, it would be interesting to highlight how international literature (academic as well as that of international organizations such as Organization for Economic Cooperation and Development) addresses urban sprawl in EU and the USA contexts and the similarities and differences among countries (Lanzani, 2012). The US experience represents a diffused phenomenon that takes root in small urban centres or in country areas with various degree of urbanization, and it is the outcome of a decentralization process of residence and work outside of the nearest urban centre. Even from a functional perspective, settlement diffusion in our regions is rather complex, also because it persists along with more recent trends, that move towards a higher specialization and some sort of comingling between residential and productive use. On the whole, the trend emerging is a distribution of settlements across the territory that can be better described as a more or less organized extension of small or medium towns rather than as the growth of a compact city.

To better represent settlement diffusion, for Tuscany we calculated another indicator drawing from more detailed information that allows us to investigate the phenomenon at sub-regional levels, for example circumscribing the analysis to the metropolitan area of central Tuscany (Figure 6).

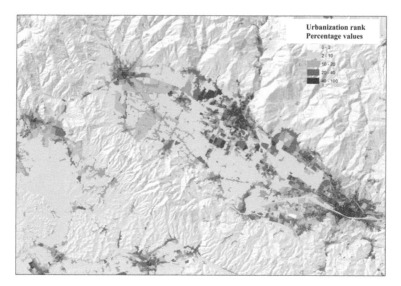

Figure 6. Saturation of the Florence–Prato–Pistoia metropolitan area.
Source: IRPET elaboration on data from the Regional technical map and ISTAT.

The indicator presented here gives us the ratio between the surface occupied by the single buildings (included in the technical map of Tuscany at the scale 1:10,000) and the surface resulting from the ISTAT 2011 census sections, thus measuring the saturation level for each section.

Obviously, the peak values are reached at the poles of the metropolitan area. Saturation reduces as we move from the historic heart of the city and, in parallel, the level of settlement diffusion increases alongside the conurbations in which the settlement fabric is more porous. If we examine more specifically the Florence metropolitan area, we notice that the connections between the two poles of Florence and Prato have higher density, while the conurbations between Prato and Pistoia are much more fragmented.

In conclusion, we can say that the present analysis allows a closer look at the actual physical combination of the poles existing in the Florence metropolitan area, and offers an account of the partial and incomplete process of metropolization that has taken place there. What we draw from it is a resilient polycentric structure, which is still visible today but, at least morphologically, seems weakened by the settlement diffusion that results from the decentralization of the areas located between the system's poles.

6. Conclusions

The examination of some second-tier cities' patterns existing in our country, and which developed out of the traditional district model, allowed the identification of different growth territorial configurations. In this work, we analysed their impact in terms of urban competitiveness and sustainability, and we found out different models of urban development. This work is based on the theoretical identification of urban functions (based on rareness) and on the straight definition of urban and regional competitiveness. Furthermore, it proposes an experimental index of sustainability in terms of land consumption and fragmentation. A first model of urban development, which characterizes Veneto and Marche, can be illustrated by the absence of stronger polarities. In fact, the first region has a diffused expansion of settlements in the territory, and the second has very weak central poles; they are both still organized as a district-like system. The other regions are characterized by stronger territorial structures, but with different levels of urban hierarchy: a stronger hierarchical structure in the case of Emilia-Romagna (which is nourished by richer economic and productive systems) and a weaker one in the case of Tuscany, which suffers from a lower ranking of both the central pole and the small–medium urban poles compared to the previous one. Regions' competitiveness is measured as their ability to "export" rare urban functions, that-is-to say to attract extra-regional demand for highly qualified functions, like health, education or business trips. The situations in which medium to high urban poles are dominant present an optimal degree of urban competitiveness and quality of the services supplied, which denote both the high and evenly distributed local demand, and the existence of central poles endowed with highly specialized and wide-ranging functions.

Areas characterized by a more organized territorial structure showed higher capability of exporting urban functions and higher territorial sustainability, due to compact urban expansion. From the sustainability viewpoint, land consumption and urban fragmentation signal a more negative impact of the diffused settlement models typical of the two regions with a weaker territorial pattern. This is mainly the result of a lack of attention to territorial plans and the planning culture in dealing with spatial dynamics, which nonetheless resulted in an absence of strong polarities and a moderate consumption of territorial resources. The cases of Emilia-Romagna and Tuscany are to be placed among the regions having a relatively better position in terms of regional growth ratio of land consumption. This does not imply that the production processes of these regions generate a "low-intensity" land consumption, but only that they benefit from relative advantages when compared to both more urban-concentrated and more widespread areas. Finally, in the best performing areas, urban growth did not result in too large a waste of territorial resources.

Consequently, the comparison of different agglomeration typologies evidences the existence of an urban hierarchy also among small–medium city regions. A better competitiveness in terms of well-functioning and sustainability was discovered to correspond to an even distribution of unique functions among poles, accompanied by an endowment of widespread rare urban functions in the central area. The pattern providing the best performances (at least concerning the aspects taken into account in this study) is organized in a set of relatively strong secondary specialized poles with a core characterized by a unique functional variety.[12] Even in a context of second-tier cities, stabilization and growth must apparently go through the consolidation of specialized functions in the small–medium cities, accompanied by the enlargement and improvement of functional variety in the central pole, which can allow for higher urban ranking, and consequently better performances for the whole region. These results are also in line with what emerged in works investigating the urban dimension at an international scale (cf. Camagni et al., 2013, 2014).

From this analysis, which has been applied to the role of the territory and the urban structure in the development paths, to the economic and territorial sustainability of second-tier cities' patterns, and to the clear-cut relevance of functional specialization and hierarchies in these contexts, some political questions immediately arise. According to the results of this study, the consolidation of growth models based on second-tier cities must necessarily make it a priority to reinforce the main urban poles and the functional specialization of the minor ones.

In fact, the function played by the reference centre of polycentric areas appears to be crucial in terms of the ability to provide rare services in the territory and to attract out-of-region demand. The most effective reinforcement and attractiveness factors to play on are the infrastructure endowment, the territorial quality and the efficiency of services.

This work does in fact show that appropriate investments and infrastructural interventions should be channelled to the central core to strengthen the super-local network of relations, as well as the factors attracting new rare functions, capable of providing qualified public services, environmental pleasantness and quality of urban life. As we have seen, the main factors of land consumption and territorial fragmentation are not found in the improvement of infrastructures or in urban concentration; on the contrary, they seem to heighten exactly in the areas where the spread of population and economic activities is more scattered and featureless.

Besides, the urban and demographic evolution witnessed in recent years did not match with the reorganization of administrative boundaries and functions, so that the lack of correspondence between the boundaries of everyday life—which raise the demand for services and functions—and the administrative boundaries, represents a long-neglected inefficiency of the governance system.

As to the minor poles, they should go through a reinforcement of specific functional specializations in order to fully express their own potentials and to consolidate their territorial role. Local competitiveness, that is the ability to attract out-of-region demand, will in fact depend on their well-established specialization of functions.

Notes

1. The definition adopted is functional to the national territory, where it identifies the present local communities (whether it is districts, cities and metropolitan areas) starting from commuters' daily travels correlated to the organization of labour markets. The basic idea, on which studies leading to the definition of an LLS (ISTAT-IRPET, 1986) are grounded, is that daily movements take place where economic and social relationships are more dense, in a space that corresponds to the community to which people belong. Although the calculation method used by ISTAT to identify daily commuting basins is not exempt from critics, and can certainly be refined (cf. Compagnucci, 2009), the idea that a good approximation for a territorial community is given by the places of daily attendance is certainly a sound concept.

2. The data on employees were drawn from the ISTAT archive Archivio Statistico Imprese Attive of business local units, 2009. Given that the public tertiary sector, especially with a high level of qualification, plays a key role in the cities, we decided to complement this archive with the data concerning the employees of universities (Source: Ministry of Education, Universities and Research) and hospitals (Source: Ministry of Health).

3. Jacobs defines a city as "a settlement that consistently generates its economic growth from its own local economy" (Jacobs, 1969, p. 262) where, in particular "much new work is added to older work and [...] this new work multiplies and diversifies a city's division of labour" (Jacobs, 1969, p. 122).

4. The other occurrences observed for the whole of Italy are the Milan area (overall population of 5,200,000 inhab.), Bergamo (875,000 inhab.) and Brescia (530,000 inhab.) for Lombardy; the Naples area (3,005,000 inhab.) for Campania; Palermo (950,000 inhab.) and Catania (697,000 inhab.) for Sicily.

5. It is reasonable to assume that the quality of health services is influenced by the pattern of the urban structure, as it implies the attainment of a minimum demand level and the activation of an array of accessory services (university research, administrative services, public transport, and so on). The ability to attract patients from outside regions, however, cannot be simply explained with the demographic size of the region or with that of the region's urban pole. In fact, among less attractive areas, we find

some densely populated and typically monocentric Southern regions (Campania, Sicily), while among the more attractive we notice regions with an average population density and a polycentric urban pattern (Emilia-Romagna and Tuscany).

6. The data take into account the participation to exhibitions, conferences, business meetings, training and refresher courses, as well as the involvement in activities of sales, installation and maintenance services and teaching.

7. To this end, we must resort to the data supplied in the Corine project. As is well know, this sort of survey presents some critical points in what concerns its scale (a module corresponds to an area of 25 ha). Nevertheless, except for the initial and still incomplete figures produced by the Italian National Observatory on Land Consumption, Corine is still today the only source available for this kind of investigation.

8. Actually, the estimate of desirable building areas should take into account not so much altimetry but steepness. Since this information is not available for all the Italian regions, we used proxy indicators, referring to the ISTAT classification of territory according to altimetry, that is hill, plain and mountain.

9. The results of a research carried out by the Planning Department of the Tuscany Region on the growth forecasts made in all of the territorial planning tools that were in force in 1987 in Tuscany, provide the following synthetic profile: 81,781,900 m^3 is the residential volume established in Tuscany's city plans; so, assuming the parameter of 120 m^3 per inhabitant, these plans are expecting to have an increase in resident population for 680,500 units, which once added to the existing 3,570,000 units make an overall population of about 4,250,000 people.

10. The urbanized areas accounted for in the index calculation are those comprised in the items 1.1.1, 1.1.2 and 1.2.1 of the Corine legend, which respectively correspond to: continuous urban fabric, discontinuous urban fabric and industrial or commercial units.

11. Urbanized cores are computed through the survey of centroids.

12. The Tuscan system undoubtedly presents a weakness since it has a smaller functional variety, or—better to say, as already noticed in previous studies carried out by IRPET—an under-endowment of manufacturing activities and related highly specialized services. The tertiary function of the city seems therefore to be unbalanced, as its services are lowly qualified and ascribable to a "banal" tertiary. Besides, being supported by a network of small–medium cities, the urban competitiveness of the Tuscan urban system is on the whole weaker than it might be expected.

References

Brezzi, M. & Veneri, P. (2014) Assessing polycentric urban systems in the OECD: Country, regional and metropolitan perspectives, *European Planning Studies*, this issue.

Burger, M., Meijers, E., Hoogerbrugge, M. & Masip Tresserra, J. (2014) Borrowed size, agglomeration shadows and cultural amenities in North-West Europe, *European Planning Studies*, this issue.

Calafati, A. G. (2012) Le città nello sviluppo economico della Terza Italia, in: A. G. Calafati (a cura di), *Le città della terza Italia: evoluzione strutturale e sviluppo economico*, pp. 69–108 (Milano: FrancoAngeli).

Camagni, R. (1993) *Economia urbana: principi e modelli teorici* (Roma: NIS).

Camagni, R. & Capello, R. (2000) Beyond optimal city size: An evaluation of alternative urban growth patterns, *Urban Studies*, 37(9), pp. 1479–1496.

Camagni, R. & Salone, C. (1993) Network urban structures in Northern Italy: Elements for a theoretical framework, *Urban Studies*, 30(6), pp. 1053–1064.

Camagni, R., Capello, R. & Caragliu, A. (2013) One or infinite optimal city sizes? In search of an equilibrium size for cities, *The Annals of Regional Science*, 51(2), pp. 309–341.

Camagni, R., Capello, R. & Caragliu, A. (2014) The rise of second-rank cities: What role for agglomeration economies? *European Planning Studies*, this issue.

Capello, R. (2002) Economie di scala e dimensione urbana: teoria ed empiria rivisitate, *Scienze Regionali*, 2(2), pp. 79–100.

Cirilli, A. & Veneri, P. (2012) Funzioni e rango delle città della Terza Italia, in: A. G. Calafati (Ed) *Le città della terza Italia: evoluzione strutturale e sviluppo economico*, pp. 253–281 (Milano: FrancoAngeli).

Compagnucci, F. (2009) *I Sistemi Locali del Lavoro nell'interpretazione dell'organizzazione territoriale: fondamenti teorici e limiti ontologici*, Quaderno di ricerca, n. 336, Università Politecnica delle Marche.

Davoudi, S. (2003) Policentricity in European spatial planning: From an analytical tool to a normative agenda, *European Planning Studies*, 11(8), pp. 979–999.

Dematteis, G. (Ed.) (1992) *Il fenomeno urbano in Italia: interpretazioni, prospettive, politiche* (Milano: FrancoAngeli).

Di Giacinto, V., Gomellini, M., Micucci, G. & Pagnini, M. (2012) *Mapping local productivity advantages in Italy: Industrial districts, cities or both?* Tema di discussione, n. 850, Banca d'Italia.

ESPON (2007) *Study of urban function ESPON project 1.4.3*. Final report, March 2007. http://www.espon.eu/export/sites/default/Documents/Projects/ESPON2006Projects/StudiesScientificSupportProjects/UrbanFunctions/fr-1.4.3_April2007-final.pdf (accessed 30 May 2013).

Florida, R. (2003) Cities and the creative class, *City & Community*, 2(1), pp. 3–19.

IRPET (2011, 2012) *Rapporto sul territorio* (Firenze: Istituto Regionale Programmazione Economica della Toscana).

Irwin, E. & Bockstael, N. (2007) The evolution of urban sprawl: Evidence of spatial heterogeneity and increasing land fragmentation, *Proceedings of the National Academy of Sciences*, 104(52), pp. 20672–20677.

ISTAT (Ed.) (2006) *Specializzazioni produttive e sviluppo locale*, in Rapporto annuale. La situazione del paese nel 2005, Roma.

ISTAT-IRPET (a cura di) (1986) *I mercati locali del lavoro in Italia* (Milano: FrancoAngeli).

Jacobs, J. (1969) *The Economy of Cities* (New York: Vintage).

Kloosterman, R. & Lambregts, B. (2001) Clustering of economics activities in polycentric urban regions: The case of Randstad, *Urban Studies*, 38(4), pp. 717–732.

Lanzani, A. (2012) *L'urbanizzazione diffusa dopo la stagione della crescita*, in: C. Papa (Ed.) *Letture di paesaggi*, pp. 223–264 (Milano: Guerini).

Meijers, E. (2007) *Synergy in Polycentric Urban Regions: Complementarity, Organising Capacity and Critical Mass* (Delft: Delft University of Technology).

Nordregio (2004) *ESPON 1.1.1 Potential for Polycentric development in Europe*, Final Project Report. Available at www.espon.eu (accessed 30 May 2013).

Parkinson, M., Meegan, R. & Karecha, J. (2014) City size and economic performance: Is bigger better, small more beautiful or middling marvellous? *European Planning Studies*, this issue.

Parr, J. B. (2004) The polycentric urban region: A closer inspection, *Regional Studies*, 38(3), pp. 231–240.

Romano, B. (2004) *Enviromental Fragmentation Tendency: The Sprawl Index* (Portugal: Atti congresso ERSA-Porto).

Romano, B. & Paolinelli, G. (2007) *L'interferenza insediativa nelle strutture ecosistemiche* (Roma: Gangemi).

Romano, B., Vaccarelli, M. & Zullo, F. (2010) *Modelli insediativi ed economia del suolo nella culture post-rurale* (Milano: FrancoAngeli).

Van Oort, F., de Geus, S. & Dogaru, T. (2014) Related variety, regional economic growth and place-based development. Strategies in urban networks in Europe, *European Planning Studies*, this issue.

Polycentric Metropolitan Development: From Structural Assessment to Processual Dimensions

RUDOLF GIFFINGER & JOHANNES SUITNER

Centre of Regional Science, Department of Spatial Planning, Vienna University of Technology, Vienna, Austria

ABSTRACT   *In this paper we aim to enhance the prevailing structural perspective on metropolization by pointing to the mutual relationship between the processes of metropolization and polycentric development. We claim that a processual view is needed to emphasize the temporal dependencies between different layers of polycentricity, and to reveal that European city–regions are situated in different stages of polycentric metropolitan development (PMD). To illustrate this empirically, we first analyse Bratislava and Vienna as two European city–regions that recently decided to jointly approach metropolitan development, while their contextual conditions and development trajectories differ significantly. It is shown upon an indicator-based analysis that the two are in different phases of the metropolization process. Confronting this evidence with stakeholder assessments of the need for strategic intervention in metropolitan development further uncovers the importance of the strategic dimension in metropolitan research. Building upon that, we conduct cluster analysis for a sample of 50 European city–regions by the same indicator set. It is shown that even this large sample of agglomerations can be grouped by different types of metropolizes, reflecting distinct effects of the metropolization process on urban-regional transformation. Hence, we conclude that a processual understanding in strategic approaches to PMD is necessary. Only if the different phases, paces, and effects of the metropolization process are taken into account, we can formulate serious recommendations for the polycentric development of distinct European urban territories. The move from structural to processual understanding is an essential foundation to learning processes for the governance of future PMD. Furthermore, the emphasis on different types of metropolizes should be taken into account in the formulation of future European policies on metropolitan development.*

1. Introduction

European cities have encountered diverse transformations within the past quarter century. First, the internationalization of trade, attended by huge economic restructurings, reshaped the economic functioning and organization of urban agglomerations across Europe. New patterns of intense migration, mobility of capital, goods and people, and ongoing technological innovation constitute a decisive development condition, namely that of globalization (Dicken, 1998; Held *et al.*, 1999). Second, the permeability of borders is being enhanced by the not yet finished process of European integration, leading to new patterns of urban and regional development in socio-demographic and economic terms (Krätke, 2007; European Commission, 2010a). And third, recurrent economic crises confront European cities and regions with new policy challenges and a break with approved methods of multi-level governance and strategic planning (Herrschel, 2009; Camagni & Capello, 2012).

European urban agglomerations are also increasingly confronted with inter-place competition for metropolitan functions that shape their urban–regional development (Kunzmann, 1996; Friedmann, 2002). Competition under conditions of supranational economic policies and regardless of the physical geographies of distance and national urban hierarchies opened up the once stable European urban system for re-positioning. Thus, European cities actively engage in capitalizing their potentials into assets and providing area-based advantages for the attraction of human and investment capital (Camagni, 2009). Depending on their competitiveness, these cities are hence on their way to becoming metropolizes. Such efforts reach far beyond administrative city boundaries, demanding new modes of metropolitan multi-level governance (Healey, 1997; Parkinson, 1997; Salet *et al.*, 2003). In this regard, the issue of polycentricity is increasingly emphasized for its contribution to integrated metropolitan development (ESPON, 2005). Consequently, European city–regions are assessed on the micro level concerning their integrated inner development, and on a macro level as concerns their embedding in transnational and global networks (ESPON, 2005).

Such assessments though, often analyse solely structural characteristics to depict the degree of polycentricity and metropolization and define recommendations for policy and planning upon that. And the concepts of metropolization and polycentricity are persistently debated as independent analytical variables or normative visions of European urban development. Even more, while academia has for long regarded metropolization as an evolutionary process in analytical terms, i.e. to understand how it evolves and why certain functions of a global economy touch down in distinct places (Sassen, 2001; Hall & Pain, 2006; Castells, 2010), the political processes inherent in strategy-building for metropolitan development and the governance of metropolization are not taken into account with equal emphasis.

Hence, in this paper we highlight the processual dimension of metropolitan development. We claim that for analysing polycentric metropolitan development (henceforth PMD), such a perspective is not only necessary to better understand the links between metropolitan and polycentric features of city–regions and the temporal dependencies between different layers of polycentricity. It also re-emphasizes the fact that despite globalized conditions for territorial development, European city–regions still encounter surrounding transformations differently (Marcuse & van Kempen, 2000; Krätke, 2007), which consequently effectuates their pace of metropolization. Considering this, we suggest intensifying the conflation of analyses of place-based evidence with analyses of distinct local development trajectories, governance processes, and assessments of PMD by relevant stakeholders.

To elucidate this claim, we start with describing PMD in theory upon recent definitions of metropolization, polycentricity, and their presumed interrelations. Next, the argument for adding a processual dimension to the structural assessments of metropolitan regions is substantiated. Empirically, we begin with conducting an analysis of Bratislava and Vienna as two cities that recently launched a common metropolitan governance initiative. With an indicator-based analysis we attempt to uncover that the two city–regions are situated in different phases of the metropolization process. Conflating this approach with a stakeholder assessment at the same time[1], we attempt to reveal the importance of analysing locally specific stakeholder attitudes, strategic considerations and perceptions of PMD to make serious recommendations for metropolitan planning. In a second step, we try to find whether the assumption of different phases of PMD can be deemed correct on the European level. Therefore, we conduct cluster analysis by the same indicators for a sample of 50 European city–regions. Herewith we plan to show that urban regions can actually be grouped by characteristics of the metropolization process, pointing at different phases of PMD. This urges European territorial policies to further increase the acknowledgement of local specificities and foster research on place-based evidence.

2. Understanding PMD

2.1. *The Concept of Metropolization*

The process of metropolization, its driving forces, as well as its impacts on urban development have been subject to intense scientific and planning discussions since the 1980s. Facing the increasing globalization of economic activities, improved ICT, and altered modes of production, distribution and consumption, metropolization soon became a dominant debate in several fields of urban studies (Friedmann, 1986; Thornley, 2000; Sassen, 2001). For European cities, the changing geopolitical contexts made this debate even more important for their development considerations. The fall of the Iron Curtain and the process of European integration brought about new opportunities and perspectives for most cities, while shaking the urban hierarchies and position of established cities within them. The newly evolved competitive markets and cooperation possibilities soon altered chances of attracting new activities, but as well increased the challenges of urban-regional governance (Giffinger, 2005; Hamilton *et al.*, 2005; Hall & Pain, 2006).

Meanwhile, the academic discussion has settled on a distinction of two major approaches to metropolization processes, as Castells (2010) elaborates. On the one hand, the primacy of a global knowledge economy is emphasized as the driving force of metropolitan growth, which is limited to a small number of powerful nodes on the global map (Hall & Pain, 2006). As a result, urban centrality in the core still exists, while being enhanced by further functions in new specialized sub-centres in the metropolitan region. And consequently, urban sprawl is increasingly replaced by the emergence of such new sub-centres. On the other hand, Castells (2010, p. 2740) argues that, "[...] the key spatial feature of the network society is the networked connection between the local and the global". In this perspective, places are connected upon their contribution to the network's quality. This contribution, again, depends to a large degree on these places' respective local networks. Hence, the process of metropolization is driven by the interaction between global and local relations. And it affects the spatial, as well as the socio-demographic development of a city–region. Both networks demand exceptional

ICT standards and transport infrastructures that facilitate global accessibility, as well as face-to-face interaction on a very local level.

From a place-based view, the varying impacts of global networks and European integration are considered as key factors enforcing the competition between cities. Hence, the European urban system experiences two decisive changes. First, not every city is able to meet the new development challenges. Economic restructuring, new economic functions, the increase of knowledge intensive activities, immigration, and the disappearance of labour intensive industries decisively affect the socio-spatial development of metropolitan regions. Herein, some cities and neighborhoods lose, while others reside as winners in economic and social terms (Fainstein *et al.*, 1992; Sassen, 2001; Krätke, 2007). Second, established cities experience particular challenges due to increased competition for their formerly distinct economic, cultural and political functions. Consequently, these functions are relocated to only a few cities globally (Sassen, 2001; Krätke, 2003; Hall & Pain, 2006; Castells, 2010). Accordingly, metropolization is predominantly driven by the allocation of such specialized functions. In order to attract them, cities need to be attractive not as single nodes, but as part of an urban region, so that these functions can be allocated to those places with the highest area-based advantages, herewith supporting the interaction of actors in global and local networks. In line with the points discussed, we can hence define the essential characteristics of the metropolization process as follows:

- Allocation of (new and specialized) economic functions and population as a factor of growth and spatial extension towards a metropolitan region (cf. for instance Friedmann, 1986, 2002; Geyer, 2002; Hall & Pain, 2006)
- Exercise of command and control functions in global networks of material and immaterial flows with excellent connectivity between urban nodes (cf. for instance, Keeling, 1995)
- Technological innovation and economic restructuring towards knowledge intensive economic activities in specialized branches of production or service (cf. for instance, Krätke, 2007; Castells, 2010)
- Socio-spatial processes of segregation or fragmentation through increasing social polarization and the replacement of old urban functions by high-ranked economic functions (cf. for instance Marcuse & van Kempen, 2000; Sassen, 2001)

A specific aspect in conceptualizing metropolization processes is the spatial de-concentration of specialized functions. It is considered important for handling growth and securing both competitiveness and territorial cohesion at once. Hence, the concept of polycentricity is recurrently mentioned as a key element of metropolitan development. A number of studies point at the increasing decentralization of metropolitan functions, identifying that they are being housed in metropolitan sub-centres (Krätke, 1995; Kunzmann, 1996, Friedmann, 2002; ESPON, 2005, 2006, 2012). Yet, as these sub-centres are the outcome of the interplay between global and local networks, metropolitan development must be interpreted as the spatial convergence of urban dimensions of multi-layered global networks. From this perspective, polycentric development is therefore a specific layer in the spatial context of metropolitan development, which replaces the urban-regional model of urban sprawl with that of morphological and functional polycentricity. Its exact definition and importance for metropolization processes are thus elaborated in the following section.

2.2. *The Concept of Polycentricity*

The development of metropolitan regions can neither be analytically explained, nor strategically approached without taking into account the obvious specificity of their spatial and functional configuration. As Roca Cladera *et al.* (2009, p. 2842) claim: " The reality of urban development from the 1980s has revealed substantial changes in the structure of metropolitan areas, which cannot be explained by the standard model." And they go on to elaborate that it is particularly the polycentric structure of these metropolitan territories that deserve our attention, as new sub-centres are increasingly found to be contributing decisively to the economic performance and stability of urban systems with metropolitan character (Riguelle *et al.*, 2007; Roca Cladera *et al.*, 2009; Camagni *et al.*, 2013).

In its simplest, polycentricity describes the circumstance that the structure and development of a metropolitan territory are determined by multiple instead of a single node (Roca Cladera *et al.*, 2009; ESPON, 2012). Today, polycentricity is debated in a multi-faceted way as both an analytical concept to reveal the level of multi-scalar integration of metropolitan urban regions (ESPON, 2012; Kramar & Kadi, 2014) and as a normative goal to alleviate the problems attending metropolization processes (Council of Ministers, 2011). The final report of ESPON 1.1.3 reveals this duality: "[P]olycentricity can be conceptualized as both an ongoing process and as a normative goal to be achieved and is alleged to help in reducing regional disparities and in increasing competitiveness for integration." (ESPON, 2006, p. 12). Yet, polycentricity is—notwithstanding its long career as a theoretically discussed and empirically applied concept—critically debated regarding its positive impacts on urban growth, territorial competitiveness and equal spatial development (Vandermotten *et al.*, 2008; Herrschel, 2009; Maier, 2009). Still though, corresponding strategies emphasize the enforcement of polycentric structures in order to enhance competitiveness and alleviate negative side effects of metropolization processes, allowing for cohesion within metropolitan territories. "[T]he polycentricity model [...] is seen by policymakers as less likely to be exclusive, because it reduces imbalances between dominant cities and 'the rest'. " (Herrschel, 2009, p. 243). Thus, polycentricity has become a widespread normative goal in metropolitan and European development strategies due to its usability regarding territorial development visions (European Commission, 2010b; Council of Ministers, 2011; ESPON, 2012).

The concept subsumes a number of facets to analytically describe the level of integration of urban agglomerations. In terms of scale, micro-, meso- and macro-levels are usually distinguished to describe the polycentric constitution of a territory. Micro-level polycentricity points at the level of internal integration of metropolitan regions, i.e. a city and its surroundings. Meso-scale polycentricity defines dense national urban networks and the occasional case of intense border-crossing relations (e.g. so-called potential polycentric integration areas, or PIAs). Slovenia's national urban network is one often-cited case of such well-integrated meso-scale polycentric networks (ESPON, 2012). Macro-scale polycentricity ultimately describes the embedding of metropolitan regions in wider transnational or global networks of metropolizes—the "Pentagon" being an example frequently referred to (ESPON, 2005, 2006).

An equally important distinction concerns the quality of polycentricity. Several accounts of PMD have already formulated a number of conceptualizations that are ever distinct, depending on the focus of analysis and available data for revealing empirical evidence of certain polycentric structures. This paper does not aim at producing a review of the exhaustive literature on theorizing polycentricity, particularly as recent scientific work has already intensely engaged in this effort (Vandermotten *et al.*, 2008; Kramar & Kadi, 2014). Instead, we build upon a differentiation of three basic types of polycentricity by condensing the manifold approaches to a workable definition for this paper and its empirical considerations[2]:

- *Morphological polycentricity*: the delineation of urban hierarchies based upon size; polycentricity as structural characteristic regardless of inter-nodal relations; place-specific city–region definitions and rank-size distribution within the metropolitan territory
- *Functional polycentricity*: the allocation of infrastructural networks, flows and inter-actions between urban nodes as indicators of inter-urban relations; technical infrastructures, distance, commuting on a daily basis, networking activities in economic, scientific and socio-cultural concerns as indicators
- *Strategic polycentricity*: the identification of political-institutional relations as signified in policy processes and strategic development documents; the cognitive envisioning of PMD as done by relevant stakeholders in strategic urban development processes; inter-urban cooperation, strategic networking between municipalities, planning agreements

Demarcating a regionally integrated metropolis with its polycentric structure is seemingly unproblematic in morphological and functional terms on a micro scale. Yet, delineation is more problematic in strategic or political terms. Diverse imaginations exist of what is to be subsumed spatially, functionally, economically, and politically under a polycentric metropolitan region throughout the range of stakeholders. Also, goal conflicts between a micro-scale delineation of a metropolis as a functionally integrated agglomeration (i.e. welfare- and inclusion-oriented) and the macro-scale imagination of a metropolis as a city in competition (competitiveness-oriented) have been long known (ESPON, 2006; Herrschel, 2009). As both morphological polycentricity and functional linkages can be decisive preconditions in the arrangement of strategic polycentricity, it is necessary that they are perceived by relevant stakeholders. Thus, analyses of recent processes of PMD should attempt to confront evidence with strategy to see how, or if at all, actual development conditions are perceived accurately, and if the respective phase and pace of PMD are taken into account by decision-makers. Hence, it is important to adopt a processual dimension in the conceptualization of PMD, where the different layers of polycentricity are understood as interrelated factors that are temporally and logically dependent on each other.

2.3. *Adopting a Processual Dimension in Analysing PMD*

The above-introduced concepts of metropolization and polycentricity are regarded as mutually related influencing factors of urban development in most scientific discourses (cf. for instance Krätke, 1995; Roca Cladera *et al.*, 2009; Castells, 2010). Yet, as recent research shows, often enough in strategic discussions and urban development practice they are rather independently approached. Also, these approaches point to the assessment of PMD from a structural perspective only, and do not sufficiently embed PMD in the individual development contexts of distinct city–regions and the different stages of metropolization, in which these city–regions can be situated (ESPON, 2012). This, although we know that urban development, and consequently also metropolitan development, is path-dependent (Moulaert & Jessop, 2013), meaning it leads to ever-specific outcomes. European city–regions have been facing different and changing geo-political and economic preconditions over the last century, hence revealing completely different paths of development based on their experiences and ability to meet new challenges (Krätke, 2007). Also, temporal and logical dependencies between different layers of polycentric development are not analysed, although being of importance as a knowledge base for implementing territorial development strategies. Hence, we suggest enhancing current approaches to PMD with a processual dimension that explicitly points to these interrelations and dependencies. Yet, how can we argue that such a processual dimension is necessary?

First, because one can assume that the metropolization process is mutually interrelated with polycentric development on both the micro- and macro-levels—a notion recurrently supported by several scholars (Kunzmann, 1996; Leroy, 2000). At least on the micro-level, metropolization processes are by their very nature dependent on the polycentric region for different reasons—for instance the attractiveness of a metropolitan region exceeding the core city, or a well-organized, decentralized region as a potential for future growth. In fact, robust polycentric structures within a metropolitan region are a positive influencing factor of metropolitan growth (Camagni et al., 2013). Consequently, cooperative polycentric strategies are needed that aim at fostering the most relevant assets on the regional level, steering competition between participating cities and municipalities (Giffinger & Hamedinger, 2009). On the macro-level we can see similar dependencies. Metropolization needs macro-polycentricity, as the specialization of metropolizes enforces functional relations with other cities and regions. Furthermore, macro- PMD is supported through the embeddedness in global networks of material and immaterial flows or the exercise of command and control functions (Keeling, 1995).

Second, because we need to consider the dependencies between the different qualities of polycentricity. Morphological polycentric features might tell us much about structural preconditions for establishing functional ties and the focus of future territorial cohesion policies. Functional polycentricity is as much a valuable account of a currently well- or scarcely integrated metropolitan region, pointing at the need for adjustment in active spatial development strategies (Geppert, 2009). Yet, we need to consider that polycentric structures in terms of morphology are the material outcome of functionally intense, long-term relations, and that these functional ties are often dependent enough on earlier strategic polycentricity endeavours regulating or facilitating these functional ties (Geppert, 2009).

Thus we argue that the chronology in PMD must as well be considered. In this regard, polycentricity is not just a matter of spatial development, but has an equally important processual dimension that needs to be considered in policy-processes. Analytical and strategic polycentricity thus need to be integrated to support territorially cohesive metropolitan development. Strategic decisions should be made on the basis of broad stakeholder integration, their perception of current polycentric structures, their visions for future PMD,

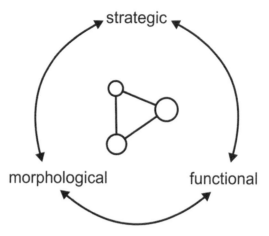

Figure 1. Processual dependencies of the different layers of polycentricity.

and empirical evidence from an analytical standpoint to conjoin all facets of polycentricity at the same time (cf. Figure 1).

Yet, of course, such a processual dimension only makes sense if we are sure that metropolization processes are in fact context-dependent and, hence, differ between European city–regions. Only if we can discover that PMD proceeds at different paces in different city–regions and that, hence, these city–regions are currently in different phases of the metropolization process, it would be helpful to point to the temporal dependencies of metropolization and the layers of polycentricity for the governance of territorial development. Otherwise we could as well turn to undifferentiated territorial policies that neglect place-based evidence and path-dependencies in planning.

Thus, we first analyse Bratislava and Vienna as an often referred-to case of cooperation in metropolitan governance to see whether the assumption of different phases of the metropolization process is correct. And we confront territorial evidence with strategic assessment in this example to show that it is important to empirically analyse the different layers: a strategic or political one vis-à-vis a functional and morphological layer. Second, we take our assumption to the European level to test whether—in the fragmented, path-dependent metropolization processes—similar types of metropolizes can be revealed that might help us to better understand the process of PMD and allow for formulating future European policies for diverse city–regions.

3. Phases of PMD: Comparing Bratislava and Vienna

Based on the above discussion it becomes obvious that metropolitan development needs to be interpreted as being deeply linked with the global embedding of metropolizes in wider functional networks, i.e. macro-polycentric development. Metropolitan development and distinct conditions for establishing and steering links with other urban nodes stimulate each other mutually. High-ranked functions and headquarters, congresses or the establishment of knowledge-intensive activities demand excellent accessibility in order to function properly. While this relationship has already been intensely debated (cf. for instance Sassen, 2001; Hall, & Pain 2006; Castells, 2010), it needs to be highlighted with equal emphasis that metropolitan development as well needs a robust micro-polycentric foundation on the urban-regional level. Urban growth and transformation processes demand polycentric structures on the urban-regional level as, according to Camagni *et al.* (2013), micro-polycentric development supports the reduction of urban costs in metropolitan development.

Hence, we can argue that metropolitan and polycentric development processes are interrelated—their form depending on the ever-specific development paths taken by a particular city–region. Correspondingly, we look at selected European metropolizes in terms of specific morphological and functional features and the strategic layer of PMD. Bratislava and Vienna are analysed in a first step as a case of cooperative metropolitan governance of two nodes in the Central European urban system. Most obviously, the two differ in terms of the preconditions to PMD already at first sight—not only in terms of size, but also as concerns their roles as capitals of Central European countries with distinct political-economic histories (Giffinger & Hamedinger, 2009).

Accordingly, an empirical analysis of the two city–regions is conducted, which groups indicators into components that can be considered characteristic of PMD. First, we regard metropolitan growth as a basic component reflecting size, current urban development process, and an agglomeration's attractivity in a competitive context. Second, we define high-ranked functions as a component of PMD, giving recognition to the importance of command and control functions and global accessibility in the metropolization process. Third, economic restructuring is reflected by indicators pointing at a transformed labour market due to a shift to a knowledge-based economy (KBE). Fourth, city–regional integration covers the micro-polycentric constitution of a metropolitan region by reflecting urban-regional structural and functional imbalances. And fifth, transnational embeddedness defines the macro-polycentric integration of a metropolitan territory in wider functional networks. Each of the five components is described by a bundle of indicators (cf. Figure 2).

The basis for our empirical analysis is the classification of city–regions according to the ESPON project 1.1.1 (ESPON, 2005). From 1595 FUAs (Functional Urban Areas) with more than 20,000 inhabitants, 50 city–regions are finally nominated as a representation of European average in terms of PMD.[3] The respective city–region's figures can then be mirrored with the sample's average to allow for classifying its individual performance. At the same time, we confront this indicator-based evidence with a qualitative assessment of PMD by relevant city–regional stakeholders in Bratislava and Vienna. This is meant to

PMD Characteristic	Component	Indicators
Population and settlement growth	**Metropolitan growth**	• Annual population growth rate (1990-2007) • Increase rate of built-up areas/capita (2000-2006)
Command and control functions *Global accessibility*	**High-ranked functions**	• No. of. headquarters of transnational firms (2006) • Accessibility of metropolitan region (2010)
Knowledge-intensive businesses and services	**Knowledge-based economy (KBE)**	• Share of population with tertiary diploma (2005) • Share of scientific and technical employment (2005)
Micro-polycentricity	**City-regional integration**	• Commuting disparities (2004-2010) • Development disparities (2004) • Population growth difference (2000-2005)
Macro-polycentricity	**Transnational embeddedness**	• No. of congresses held in region (2009) • No. of air passenger (2006) • Share of Erasmus students (2008-2009)

Figure 2. Defining components of PMD (see appendix).

elucidate to what degree certain preconditions of PMD are reflected in stakeholder's perceptions, thereby bringing in the strategic dimension of polycentricity, which—as elaborated above—is important to fully comprehend the evolutionary process of metropolitan development that is currently still under-represented in European regional and planning studies.

This first analysis draws upon results from research conducted within POLYCE, a European research project that was run from 2010 to 2012 within the ESPON 2006–2013 Programme (ESPON, 2012). Within POLYCE, metropolization processes and polycentric development of five Central European capital cities, namely Bratislava, Budapest, Ljubljana, Prague and Vienna, were analysed upon European territorial indicators. As another integral step, focused workshops were organized in all five capital cities, with 20–30 relevant stakeholders from the respective metropolitan regions attending. POLYCE created Metropolitan Agendas as development strategies upon stakeholders' opinions, ideas and visions about the future development options for the five Central European metropolitan regions. Participating actors were asked to contribute to the creation of these agendas with their expertise by sharing their perception of recent PMD processes and assessing current development paths of their respective metropolitan region. The qualitative judgment deriving from this strategic discussion was then conflated in a bottom-up manner by the research team and clustered thematically, the result being a Metropolitan Development Agenda for each of the city–regions. Confronting this qualitative assessment with the above introduced indicator-based components reveals in which fields of intervention strategic actors of well- or poorly embedded metropolizes see a need for action, which are less regarded, and to what degree these interpretations converge with the evidence provided above.

Figure 3 shows the results. Territorial evidence, i.e. the components of PMD, and the strategic discussion, i.e. the respective qualitative assessments, were normalized to obtain comparable figures. For territorial evidence, a positive deviation of one PMD component from the European average is interpreted as a low need for strategic intervention, while a negative deviation indicates the opposite. Concerning the results from the strategic stakeholder discussions, the number of mentions of activities relating to a metropolization and/or polycentricity feature is grouped by similar categories. "Low" stands for only few

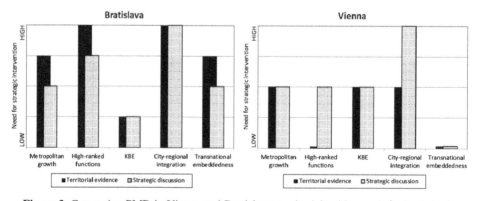

Figure 3. Comparing PMD in Vienna and Bratislava: territorial evidence vis-à-vis strategic assessment.

related activities, while "high" indicates a high number of mentions and, thus, an awareness of the need for strategically intervening in the PMD path of the respective metroregion. Hence, this classification allows for a simple comparison of territorial evidence (i.e. the descriptive analysis of indicators compared to European average) and strategic discussion (i.e. the qualitative assessment of activities mentioned in the Metropolitan Agendas of the Central European metropolitan regions).

Obviously, the two compared city–regions differ decisively as concerns the indicator-based analysis of PMD. Whereas Vienna performs largely above average in most of the defined components, Bratislava's results vary largely across the five categories. Results particularly hint at Vienna's role in a wider urban system, where it is obviously well-embedded. Therefore, the strategic debate particularly concentrates on fostering city–regional integration processes instead of further pushing processes of metropolization. Stakeholders seem to accurately interpret the development conditions of the Vienna metropolitan region, while overemphasizing only the equipment with command and control functions and the need for further urban-regional cohesion. Bratislava instead shows a more dispersed picture concerning the degree of convergence between evidence and strategy. Particularly in terms of metropolization features, actors give less regard to activities fostering their metropolitan development than would actually be needed considering indicator analysis. Although tendencies here also point at a convergence between both evidence and strategy, the emphasis on strengthening growth, high-ranked functions and global embedding should be higher in all four fields. As concerns city–regional integration on the other hand, actors already seem to be aware of the high demand for stronger micro-polycentric development efforts. In general, the need to catch up in European terms is by and large acknowledged, although suggested activities cannot cover the whole range of interventions needed.

This initial analysis of two metropolitan regions hence uncovered two things: first, that the structural conditions for PMD differ decisively in the two metropolizes, herewith emphasizing the need for place-based evidence as a foundation of territorial policies; and second, that despite evident structural development conditions, stakeholder perceptions in the strategic debate on PMD are equally important influencing factors of the metropolization process. The latter particularly demands reconsidering analyses of PMD as an evolutionary process depending on the planning-political (or strategic) layer of polycentricity.

4. Identifying types of PMD

Metropolization is regarded as a global process affecting urban development in different dimensions. Hence, it is reasonable to assume that it can be described empirically by a certain set of indicators for any sample of cities. According to the above theory-led discussion and the insights gained from the analysis of Vienna and Bratislava, these components reflect a city's situation in the process of metropolization in a multi-dimensional way. At the same time, contextual preconditions like the persistence of built urban structures or restricted abilities to govern urban development trends do not have the same impacts on PMD in each and every city (cf. Chapter 2.3, resp. Friedrichs, 1985, or Hamilton et al., 2005). Consequently, we can assume that cities experience the process of PMD differently. They are the result of a specific socio-political context and of a specific pace of development. Based on the above defined components we therefore classify European metropolizes by elaborating comparable features as the path-dependent outcome of recent trends of metropolization. Also, we substantiate the above empirical analysis of Bratislava and Vienna by taking the assumption of PMD as a path-dependent process to the European level. According to the above definition, the empirical classification of metropolizes will concentrate on non-correlating indicators that describe the metropolization process of European urban regions. These are the same that constitute the above defined components (cf. Chapter 3).

The five components describing PMD are standardized and the Ward-method as a hierarchical classification procedure is applied in order to identify homogenous groups of metropolizes. For figuring out the number of clusters with the most homogenous result, a repeated analysis showed that five clusters deliver a satisfying result in statistical terms. Clustering reveals five groups that differ strongly among the five components of PMD. Only two components ("City-regional integration" and "Transnational embeddedness") show insufficient internal homogeneity in comparison to the total distribution of values. Figure 4 shows the results of cluster analysis, describing the five types compared to European average.

Cluster 1: *Established metropolizes with excellent macro-polycentric performance*
City–regions in this group grow moderately, while clearly lagging behind in terms of the KBE. The latter is expressed by the lowest average among the five clusters. To the contrary, this cluster is characterized by above average "High-ranked functions", indicating their importance as global political, cultural, and economic centres. Moreover, they are characterized by a moderate decrease in "City-regional integration"—an indication of under-developed micro-polycentricity— as well, while at the same time being excellently embedded on the global scale.

Members of this cluster are: Vienna, Prague, Barcelona, Milano, Roma, La Valletta, and Manchester

		Metropolitan growth (Z-value)	High-ranked functions (Z-value)	Knowledge-based economy (KBE) (Z-value)	City-regional integration (Z-value)	Transnational embeddedness (Z-value)
Cluster 1	Avg.	-0,19	0,36	-0,58	-0,16	1,20
	N	7	7	7	7	7
	Var.	0,21	0,65	0,31	0,40	0,40
Cluster 2	Avg.	-0,60	-0,80	-0,27	1,28	-0,62
	N	11	11	11	11	11
	Var.	0,49	0,10	1,16	0,35	0,32
Cluster 3	Avg.	-0,10	1,36	1,26	0,01	0,87
	N	9	9	9	9	9
	Var.	0,58	0,99	0,42	0,21	0,27
Cluster 4	Avg.	-0,24	-0,25	-0,20	-0,81	-0,78
	N	17	17	17	17	17
	Var.	0,19	0,40	0,73	0,35	0,15
Cluster 5	Avg.	2,16	-0,28	-0,15	0,11	0,64
	N	6	6	6	6	6
	Var.	0,36	0,92	0,48	1,28	1,01

Figure 4. Clustering metropolizes by components of PMD indicators.[4]

Cluster 2: *Metropolizes with under-developed polycentric features*
City–regions of this group show the weakest performance in "Metropolitan growth" and "High-ranked functions". Even the performance in terms of KBE is slightly below European average. Hence, these city–regions face clear deficits in the process of metropolization. Besides, they are characterized by weak "City-regional integration", i.e. mono- instead of polycentric structures, and very poor "Transnational embeddedness", which reveals their weak preconditions in terms of macro-polycentric development.

Members of this cluster are: Budapest, Bratislava, Ljubljana, Sofia, Tallinn, Vilnius, Riga, Warszawa, Porto, Lisbon and Bucharest

Cluster 3: *Restructured metropolizes with advanced micro- and macro-polycentricity*
This cluster is defined by moderate "Metropolitan growth" and the best performance concerning "High-ranked functions" and KBE, expressing these city–regions' advanced status in terms of PMD. This is accompanied by a well-balanced "City-regional integration" as an indication of a balanced micro-polycentric development, and excellent "Transnational embeddedness", which fosters their macro-economic polycentric development.

Members of this cluster are: Brussels, Munich, Berlin, Frankfurt, Copenhagen, Helsinki, Amsterdam, Stockholm and Glasgow

Cluster 4: *Moderately restructured metropolizes with less advanced micro- and macro-polycentricity*
The process of metropolization did not affect these cities with equal intensity. "Metropolitan growth", "High-ranked functions", and KBE all reveal a comparably poor performance below European average. At the same time, "City-regional integration" and "Transnational embeddedness" show very strong deviations from European average, indicating a non-balanced micro-polycentric development and poor condition for macro-polycentricity.

Members of this cluster are: Antwerp, Stuttgart, Bremen, Hamburg, Düsseldorf, Köln, Lille, Bordeaux, Lyon, Torino, Bologna, Luxemburg, Rotterdam, Lodz, Krakow, Gdansk and Malmö

Cluster 5: *Fast growing, well-embedded metropolizes with inhomogeneous polycentric features*
City–regions in this cluster are fast-growing, performing moderately in terms of "High-ranked functions" and KBE. The process of metropolization is, in this group, supported by a well-balanced "City-regional integration" in terms of micro-polycentric development and above average "Transnational embeddedness" as an indication of well-developed macro-polycentric features.

Members of this cluster are: Madrid, Valencia, Seville, Toulouse, Athens and Dublin

To sum up, statistical analysis shows that the process of metropolization—in combination with polycentric features—actually allows for the identification of comparable city–regions in five homogenous groups or types of metropolizes. This classification clearly shows the very specific impact of recent trends of globalization and economic restructuring on PMD under the respective local conditions. Importantly, the above discussed and assumed mutual interrelation between certain features of metropolitan and polycentric development exists in specific bundles of characteristics that together form a characteristic cluster. However, the interrelated development process works differently for specific features and evidently provides different types of metropolizes. Hence, a general trend and interrelation that would be significant across all types of metropolizes could not be detected. This, of course, should have strong impacts on metropolitan policy approaches, which should acknowledge this fact.

5. Conclusions: From Structural Assessment to Processual Dimension

This paper concentrated on the multi-faceted understanding of polycentric development and its significance for European metropolizes. Two different empirical approaches were applied to reveal the preconditions of polycentric development in metropolitan regions. We started with an outline of the concepts of metropolization and polycentricity as our theoretical grounding. It was made clear that metropolization needs to be considered as a multi-dimensional concept that cannot be seriously discussed without taking the specific characteristics of polycentric development processes into account. For polycentricity, we highlighted three conceptual qualities: morphology as an indication of spatial structures, functional relations as the collective name for all kinds of flows and boundary-crossing ties, and strategic interests as the political dimension of PMD.

Building upon this theoretical debate and the so-produced characterization of both concepts, we then attempted to introduce two empirical approaches to identifying different phases and paces of PMD and different types of European metropolizes. We used both a mixed-methods approach and a statistical approach to point at the presumed mutual relation and the strategic dimension of processes of metropolization and polycentric development. In a first step, we concentrated on the specific situation of PMD by comparing Vienna and Bratislava. Analysing the two with a mixed-methods approach that would confront descriptive data analysis and a perceptive-assessing approach, we were not only able to uncover that the two city–regions are situated in different phases of the metropolization process, but that the local conditions for future PMD are also differently perceived among local stakeholders here and there. Obviously strategic actors in already well-integrated city–regions are aware of the importance of micro-polycentricity for PMD, although city–regional relations are perceived differently in each case. The general tendency found is that less-established metropolizes are clearly focusing a competitive behaviour, which puts core cities at centre stage, while established metro-regions have already gone one step further in their efforts to govern the metropolization process by putting more emphasis on embedding an integrated metropolitan region in a wider urban system. Still though, CE metropolizes are obviously aware of the importance of polycentricity in metropolization processes, as two thirds of suggested measures clearly aim at fostering polycentric structures. However, a comprehensive research across all 50 European metropolizes would be necessary in order to clarify further interrelations between place-based evidence and stakeholders' perceptions. Anyhow, we highlighted the importance of the strategic dimension in PMD processes only upon the basic comparative analysis of these two cases.

In a second step, we explored whether specific types of metropolizes could be recognized among a sample of 50 European urban agglomerations. We considered this analysis as an explorative approach based upon theory-led arguments. Herewith we wanted to point at the need to consider the place-based particularities of different metropolizes in policy processes of European urban development. Statistical analysis found five types of metropolitan agglomerations: first, established metropolizes with excellent macro-polycentric performance, second, metropolizes with under-developed polycentric features, third, restructured metropolizes with advanced micro- and macro-polycentricity, fourth, moderately restructured metropolizes with less advanced micro- and macro-polycentricity, and fifth, fast growing, well-embedded metropolizes with inhomogeneous polycentric features. These types were identified in terms of their specific characteristics of PMD on the micro- and macro-levels. Hence, empirical analysis has also clearly pointed to the interrelation between specific metropolizes and their polycentric structures, while it also made clear that European metropolizes are too different to make any general statements.

Results, therefore, show that actually urban regions are affected by metropolization in ever-specific ways that can still be grouped by homogenous features. Facing this fact, we urge that European territorial policies should further increase the acknowledgement of local specificities and foster research on place-based evidence. Moreover, results demand that future research on PMD take on a processual dimension that recognizes the temporal dependencies of different qualities of territorial development as well as the locally specific development trajectories of European city–regions. In this regard, further comprehensive evidence-based approaches are strongly recommended for analysing metropolitan development processes of European urban agglomerations. Yet, they need to be blended with city-specific approaches that are capable of taking into account place-based specificities of a metropolitan region. As our analysis has shown, suggestions of stakeholders on strategic efforts concerning future PMD do not always converge with deficits or assets of their cities. This divergence between perception and evident status is strongest where metropolization processes are least advanced. Therefore, any concept of European urban policy should emphasize and prioritize place-based research and strategies in order to meet specific local challenges when it comes to fostering metropolization processes and micro-polycentric structures through corresponding cooperative planning efforts.

Notes

1. Assessments of PMD in Bratislava and Vienna were ascertained in stakeholder workshops in the course of the ESPON project POLYCE (cf. ESPON, 2012).
2. The basic delineation builds upon the following sources: ESPON (2005, 2006, 2012), Geppert (2009), Maier (2009), and Kramar and Kadi (2014).
3. 76 European MEGAs were selected that are covered by Urban Audit for both the Core City and the Larger Urban Zone or approximations of these by NUTS regional level as defined by the ESPON project FOCI (ESPON, 2009). 69 MEGAs remained for data collection after removing huge MEGAs and those agglomerations not included in FOCI from the sample. According to the Urban Audit definitions and database coverage for 1999–2008, Large Urban Zones were used as the primary proxy for the metropolitan regions. In other cases, data were collected from Eurostat or other European research projects, approximating the metropolis by NUTS regional data. The 160 indicators retrieved allowed for a further reduction of the city sample to 50 European MEGAs that were sufficiently covered by data in 123 indicators.
4. The description is based on F-values defined through the relation of variance between group specific and total value. As all indicators are showing standardized values, the value of information of an indicator is expressed through the indicator's value of variance within a cluster: the lower the value of group specific variance in comparison to total variance of 1.00 is, the stronger an indicator's mean value characterizes the homogeneity of a group of metropolizes in a certain cluster/type.

References

Camagni, R. (2009) Territorial capital and regional development, in: R. Capello & P. Nijkamp (Eds) *Handbook of Regional Growth and Development Theories*, pp. 118–132 (Northampton: Edward Elgar Publishing).

Camagni, R. & Capello, R. (2012) Globalization and economic crisis: How will the future of European regions look, in: R. Capello & T. P. Dentinho (Eds) *Globalization Trends and Regional Development. Dynamics of FDI and Human Capital Flows*, pp. 36–59 (Cheltenham: Edward Elgar).

Camagni, R., Capello, R. & Caragliu, A. (2013) One or infinite optimal city sizes? In search of an equilibrium size for cities, *The Annals of Regional Science*, 51(2), pp. 309–341.

Castells, M. (2010) Globalisation networking urbanisation: Reflections on the spatial dynamics of the information age, *Urban Studies*, 47(13), pp. 2737–2745.

Council of Ministers (2011) *Territorial Agenda of the European Union 2020: Towards an Inclusive Smart and Sustainable Europe of Diverse Regions*. Available at http://www.eu2011.hu/files/bveu/documents/TA2020.pdf (accessed 21 October 2013).

Dicken, P. (1998) *Global Shift: Mapping the Changing Contours of the World Economy* (London: Sage).

ESPON (2005) *Potentials for Polycentric Development in Europe*. ESPON project 1.1.1 final report. Available at http://www.espon.eu/mmp/online/website/content/projects/259/648/file_1174/fr-1.1.1_revised-full.pdf (accessed 29 May 2013).

ESPON (2006) *Enlargement of the European Union and the Wider European Perspective as Regards its Polycentric Spatial Structure*. ESPON project 1.1.3 part 1, summary. Available at http://www.espon.eu/export/sites/default/Documents/Projects/ESPON2006Projects/ThematicProjects/EnlargementPolycentrism/full_revised_version_113.pdf (accessed 29 May 2013).

ESPON (2009) *Future Orientations for Cities*. ESPON Applied research project. Available at http://www.espon.eu/export/sites/default/Documents/Projects/AppliedResearch/FOCI/FOCI_final_report_20110111.pdf (accessed 29 May 2013).

ESPON (2012) *Metropolisation and Polycentric Development in Central Europe*. ESPON Targeted Analysis. Available at http://www.espon.eu/export/sites/default/Documents/Projects/TargetedAnalyses/POLYCE/FR/POLYCE_FINAL_MAINREPORT.pdf (accessed 29 May 2013).

European Commission (2010a) *Investing in Europe's Future. Fifth Report on Economic Social and Territorial Cohesion*. Available at http://ec.europa.eu/regional_policy/sources/docoffic/official/reports/cohesion5/pdf/5cr_en.pdf (accessed 21 October 2013).

European Commission (2010b) *EUROPE 2020. A European Strategy for Smart Sustainable and Inclusive Growth*. Available at http://ec.europa.eu/eu2020/pdf/COMPLET%20EN%20BARROSO%20%20%20007%20-%20Europe%202020%20-%20EN%20version.pdf (accessed 21 October 2013).

Fainstein, S. S., Gordon, I. & Harloe, M. (1992) *Divided Cities: New York and London in the Contemporary World* (Cambridge: Blackwell).

Friedmann, J. (1986) The world city hypothesis, *Development and Change*, 17(1), pp. 69–83.

Friedmann, J. (2002) *The Prospect of Cities* (Minneapolis: University of Minnesota Press).

Friedrichs, J. (Ed.) (1985) *Stadtentwicklung in West-und Osteuropa* (Berlin: de Gruyter).

Geppert, A. (2009) Polycentricity: Can we make it happen? From a concept to its implementation, *Urban Research & Practice*, 2(3), pp. 251–268.

Geyer, H. S. (2002) *International Handbook of Urban Systems* (Cheltenham: Edward Elgar Publishing).

Giffinger, R. (Ed.) (2005) *Competition between Cities in Central Europe: Opportunities and Risks of Cooperation* (Bratislava: Road).

Giffinger, R. & Hamedinger, A. (2009) Metropolitan competitiveness reconsidered: The role of territorial capital and metropolitan governance, *Terra Spectra*, 20(1), pp. 3–13.

Hall, P. & Pain, K. (Eds) (2006) *The Polycentric Metropolis: Learning from Mega-city Regions in Europe* (London: Earthscan).

Hamilton, I., Dimitrovska Andrews, K. & Pichler-Milanovic, N. (2005) *Transformation of Cities in Central and Eastern Europe: Towards Globalization* (Tokyo: United Nations University Press).

Healey, P. (1997) *Collaborative Planning: Shaping Places in Fragmented Societies* (London: Macmillan).

Held, D., McGrew, A., Goldblatt, D. & Perraton, J. (1999) *Global Transformations. Politics Economies and Culture* (Cambridge: Polity Press).

Herrschel, T. (2009) City regions polycentricity and the construction of peripheralities through governance, *Urban Research & Practice*, 2(3), pp. 240–250.

Keeling, D. (1995) Transport and the world city paradigm, in: P. L. Knox & P. J. Taylor (Eds) *World Cities in a World System*, pp. 115–131 (Cambridge: Cambridge University Press).

Kramar, H. & Kadi, J. (2014) Polycentric city networks in CEE: Existing concepts and empirical findings, *Geographia Polonica*, 86(3), pp. 137–152.

Krätke, S. (1995) *Stadt-Raum-Ökonomie* (Basel: Birkhäuser).

Krätke, S. (2003) Global media cities in a world-wide urban network, *European Planning Studies*, 11(6), pp. 605–628.

Krätke, S. (2007) *Europas Stadtsystem Zwischen Metropolisierung und Globalisierung. Profile und Entwicklungspfade der Großstadtregionen Europas im Strukturwandel zur wissensintensiven Wirtschaft* (Berlin: Lit-Verlag).

Kunzmann, K. (1996) Europe-megalopolis or themepark Europe? Scenarios for European spatial development, *International Planning Studies*, 1(2), pp. 143–163.

Leroy, S. (2000) Sémantiques de la metropolisation, *L'Espace géographique*, 29(1), pp. 78–86.

Maier, K. (2009) Polycentric development in the spatial development policy of the Czech Republic. *Urban Research & Practice*, 2(3), pp. 319–331.

Marcuse, P. & van Kempen, R. (Eds) (2000) *Globalizing Cities: A New Spatial Order?* (Oxford: Blackwell).

Moulaert, F. & Jessop, B. (2013) Theoretical foundations for the analysis of socio-economic development in space, in: F. Martinelli, F. Moulaert & A. Novy (Eds) *Urban and Regional Development Trajectories in Contemporary Capitalism*, pp. 18–44 (London: Routledge).

Parkinson, M. (1997) The rise of the European entrepreneurial city, in: *East-West Conference: Financing of Cities and Regions: Subsidiarity and Finance Potentials*, Conference proceedings, October 1996, Munich, pp. 125–136.

Riguelle, F., Thomas, I. & Verhetsel, A. (2007) Measuring urban polycentrism: A European case study and its implications, *Journal of Economic Geography*, 7(29), pp. 193–215.

Roca Cladera, J., Duarte, C. R. M. & Moix, M. (2009) Urban structure and polycentrism: Towards a redefinition of the sub-centre concept, *Urban Studies*, 46(13), pp. 2841–2868.

Salet, W., Thornley, A. & Kreukels, A. (Eds) (2003) *Metropolitan Governance and Spatial Planning* (London: Spon Press).

Sassen, S. (2001) *The Global City: New York London Tokyo* (Princeton, NJ: Princeton University Press).

Thornley, A. (2000) Strategic planning in the face of urban competition, in: W. Salet & A. Faludi (Eds) *The Revival of Strategic Spatial Planning, Proceedings of Colloquium*, pp. 39–52 (Amsterdam: Royal Netherlands Academy of Arts and Sciences).

Vandermotten, C., Halbert, L., Roelandts, M. & Cornut, P. (2008) European planning and the polycentric consensus: Wishful thinking? *Regional Studies*, 42(8), pp. 1205–1277.

Appendix

Indicator name	Description	Reference year	Spatial reference	Source
Annual population growth rate (LUZ)	Average annual population growth rate of city–regional agglomeration	1990–2007	Larger urban zone	Urban audit
Increase rate of built-up areas/capita (LUZ)	Average increase of built-up areas per inhabitant in the city–regional agglomeration	2000–2006	Larger urban zone	Urban audit

(Continued)

Appendix. Continued

Indicator name	Description	Reference year	Spatial reference	Source
No. of. headquarters of transnational firms	Total number of headquarters of transnational firms of the 2000 biggest world firms located in the city–regional agglomeration	2006	Larger urban zone/NUTS 3	ESPON FOCI project/ Forbes
Accessibility of metropolitan region	Number of metropolitan growth areas reachable by rail, air and intermodal return trips	2010	Functional urban area	ESPON FOCI project/OAG
Share of population with tertiary diploma	Proportion of the resident population aged 15 and above qualified at levels 5–6 ISCED in the metropolitan region	2005	Larger urban zone/NUTS 2	ESPON ATTREG project/ ESPON DEMIFER project
Share of scientific and technical employment	Share of people employed in scientific and technical jobs from total employment in metropolitan region	2005	Larger urban zone/NUTS 2	Eurostat
Commuting disparities (inbound—outbound commuters)	Absolute difference between inbound commuters (from metro-region to core city) and outbound commuters (from core city to metro-region)	2004–2010	Core city	Urban audit
Development disparities (CC—MR)	Disparities in the GDP-per-capita-level between the metropolitan area and its regional hinterland	2004	Metropolitan growth area	ESPON FOCI project
Population growth difference (CC—MR)	Difference between annual population growth rates of core city and metropolitan region	2000–2005	Core city/ larger urban zone	ESPON FOCI project (urban audit)
No. of congresses held in region	Total number of congresses held in the metropolitan region in the reference year	2009	NUTS 2	ESPON ATTREG project/ICCA
No. of air passenger	Total number of air passengers (embarkation and disembarkation) in the reference year	2006	NUTS 2	Eurostat
Share of Erasmus students	Share of Erasmus students per 1000 students enrolled at local universities	2008–2009	NUTS 2	ESPON ATTREG project

Index